ACSP · Analog Circuits and Signal Processing

The *Analog Circuits and Signal Processing* book series, formerly known as the *Kluwer International Series in Engineering and Computer Science*, is a high level academic and professional series publishing research on the design and applications of analog integrated circuits and signal processing circuits and systems. Typically per year we publish between 5-15 research monographs, professional books, handbooks, and edited volumes with worldwide distribution to engineers, researchers, educators, and libraries. The book series promotes and expedites the dissemination of new research results and tutorial views in the analog field. There is an exciting and large volume of research activity in the field worldwide. Researchers are striving to bridge the gap between classical analog work and recent advances in very large scale integration (VLSI) technologies with improved analog capabilities. Analog VLSI has been recognized as a major technology for future information processing. Analog work is showing signs of dramatic changes with emphasis on interdisciplinary research efforts combining device/circuit/technology issues. Consequently, new design concepts, strategies and design tools are being unveiled. Topics of interest include: Analog Interface Circuits and Systems; Data converters; Active-RC, switched-capacitor and continuous-time integrated filters; Mixed analog/digital VLSI; Simulation and modeling, mixed-mode simulation; Analog nonlinear and computational circuits and signal processing; Analog Artificial Neural Networks/Artificial Intelligence; Current-mode Signal Processing; Computer-Aided Design (CAD) tools; Analog Design in emerging technologies (Scalable CMOS, BiCMOS, GaAs, heterojunction and floating gate technologies, etc.); Analog Design for Test; Integrated sensors and actuators; Analog Design Automation/Knowledge-based Systems; Analog VLSI cell libraries; Analog product development; RF Front ends, Wireless communications and Microwave Circuits; Analog behavioral modeling, Analog HDL.

More information about this series at https://link.springer.com/bookseries/7381

Sining Pan • Kofi A. A. Makinwa

Resistor-based Temperature Sensors in CMOS Technology

Sining Pan
Delft University of Technology
Delft, The Netherlands

Kofi A. A. Makinwa
Delft University of Technology
Delft, The Netherlands

ISSN 1872-082X ISSN 2197-1854 (electronic)
ACSP · Analog Circuits and Signal Processing
ISBN 978-3-030-95286-0 ISBN 978-3-030-95284-6 (eBook)
https://doi.org/10.1007/978-3-030-95284-6

This Springer imprint is published by the registered company Springer Nature Switzerland AG
The registered company address is: Gewerbestrasse 11, 6330 Cham, Switzerland

Acknowledgments

Dec 2021, Delft

This book started life as a PhD thesis written at the Electronic Instrumentation (EI) Laboratory, TU Delft, where I have spent over 6 years including my MSc period. It is my fortune to have the chance of studying in this world-class laboratory in the field of circuit design. In this section, I would like to express my sincere gratitude to all the people from whom I received help and support. Without them, this work would not have come into existence.

My first and foremost appreciation would definitely go to my supervisor, Prof. Kofi Makinwa, for all his guidance, encouragement, criticism, and support. I entered the EI lab in 2015 as an MSc student, or a beginner in circuit design. To be honest, Kofi's high standards made the first months bitter, and I felt mentally stressed almost every time I received his email. I even thought about quitting the MSc project when the first tape-out failed to work properly. Fortunately, I managed to survive his tough training, and have been greatly influenced by his insightful and powerful way of analyzing and designing circuits. The 4 years of my PhD research journey went extremely smooth, and I find it hard to imagine how it could be made better.

I am also very grateful to many other people I received guidance from, for kindly answering all types of simple and sometimes naïve questions at the starting phase of my research. My special thanks go to Saleh Heidary Shalmany and Hui Jiang, and I would also sincerely thank Ugur Sonmez, Long Xu, and Junfeng Jiang for their patient guidance.

The EI lab is a great team, and I feel so lucky to be able to work with all its members. It is my honor to co-author chapters with Hui, Cagri, Yanquan, Saleh, Lorenzo, Fabio, Jan, and Matheus. Also, it was my pleasure to discuss technical problems with Prof. Johan Huijsing, Michiel, Qinwen, Chao, Zhong, Teruki, Guijie, Burak, Bahman, Thije, Shoubhik, Huajun, Roger, Authur, Amir, Eren, Efraim, Masoud, Sijun, Xinling, Said, Javad, Jieyu, Qi, Guangqian, Valaria, Daguang, Miao, Nandor, and Milos. I would like to thank Zu-yao, Lukasz, Ron, Ger, and Jeroen, for chip-bonding, PCB designing, and instrument maintaining. Thanks also go to my present and former roommates: Annemarijn, Jaekyum, Johan, and Amir, for all the pleasant casual chats. Besides, I want to thank our secretary Joyce for all her support. As for the thesis preparation, I must thank Sarah, Zhong, Huajun, Roger, and Burak, for their helpful reviews and valuable comments. Also, I would like to thank Jan/Thije for helping with the Dutch translation of the summary/propositions. I am also very grateful to all the thesis committee members, their nice feedback helped to improve the quality of this work.

Apart from all the colleagues mentioned above, I want to thank my friends who helped me and made my stay in Delft enjoyable: Mingliang, Yixuan, Dongbin, Weichen, Ziyu, Hong, Bo, Zhao, Zeyu, Weihan, Xiaoliang, Fei, and Yu. I had great times with you during non-working hours.

And I am really grateful to receive strong recommendations from many professors during award application and job-seeking processes, including but not limited to Prof. Nan Sun, Prof. Guoqi Zhang, Prof. Youngcheol Chae, Prof. Tsung-Hsien Lin, Prof. Huangqiang Wu, and Prof. Zhihua Wang. Such recommendations really helped as I am entering the next stage of my academic career.

Finally, I would express my deepest gratitude to my parents, who always stand by my side and support me in various ways.

Sining Pan

Contents

About the Author

Sining Pan received his BSc degree in electronic engineering from Tsinghua University, Beijing, China, in 2013, and his MSc and PhD degrees in electrical engineering (both cum laude) from Delft University of Technology, Delft, the Netherlands, in 2016 and 2021, respectively. He was a postdoctoral researcher at Electronic Instrumentation Laboratory, Delft University of Technology. In 2022, he joined Tsinghua University as an assistant professor. His research interests include smart sensors, CMOS frequency references, and $\Delta\Sigma$ modulators. Dr. Pan was a recipient of the ADI outstanding student designer award (2019) and the IEEE SSCS predoctoral achievement award (2020). He serves as a reviewer for *JSSC*, *TCAS-I*, *TCAS-II*, *TIM*, *Sensors J.*, and *T-VLSI*.

Kofi A. A. Makinwa received his BSc and MSc degrees from Obafemi Awolowo University, Ife, Nigeria, in 1985 and 1988, respectively; his MEE degree from Philips International Institute, Eindhoven, the Netherlands, in 1989; and his PhD degree from Delft University of Technology, Delft, the Netherlands, in 2004. From 1989 to 1999, he was a research scientist with Philips Research Laboratories, Eindhoven, the Netherlands. In 1999, he joined Delft University of Technology, where he is currently a full Professor and the head of the Microelectronics Department. His research interests include the design of mixed-signal circuits, sensor interfaces, and smart sensors. Dr. Makinwa has served on the program committees of several IEEE conferences, and he was the Analog Subcom chair of the International Solid-State Circuits Conference (ISSCC). He has also served the Solid-State Circuits Society as a distinguished lecturer and as a member of its Adcom. He is a co-organizer of the Advances in Analog Circuit Design (AACD) workshop and the Sensor Interfaces Meeting (SIM). Dr. Makinwa is an ISSCC top-10 contributor, and a co-recipient of several best paper awards. He is an IEEE fellow and a member of the Royal Netherlands Academy of Arts and Sciences.

Chapter 1
Introduction

Temperature plays an essential role in many physical, chemical, and biological processes. Therefore, temperature sensors are widely used for their monitoring and control. Traditionally, temperature sensors are based on discrete components, such as thermistors [1, 2] or thermocouples [3]. In the last few decades, however, smart temperature sensors, that is, integrated temperature sensors with on-chip readout circuits and digital outputs, have become increasingly popular due to their low cost, small size, and ease of use [4].

This book describes the design of smart temperature sensors for a specific application, the temperature compensation of frequency references [5–10], which demands both high resolution and high energy efficiency. By using on-chip resistors as sensing elements, sensors with state-of-the-art resolution and energy efficiency were realized. Moreover, these designs achieved competitive performance in various other aspects, such as accuracy, supply sensitivity, and chip area.

This chapter is an introduction to this book. It starts by discussing some general aspects of integrated temperature sensors, such as their applications and specifications. Then the specific challenges associated with the temperature compensation of frequency references are presented. This is followed by an introduction and comparison of the various temperature sensing elements available in CMOS technology, which leads to the choice for on-chip resistors. This chapter ends with an overview of the targeted goals and book organization.

1.1 Temperature Sensor Applications and Specifications

Smart temperature sensors can be used in numerous applications, which results in a wide variety of specifications. For example, low power consumption is a key requirement for sensors intended for use in radio frequency identification (RFID)

S. Pan, K. A. A. Makinwa, *Resistor-based Temperature Sensors in CMOS Technology*, ACSP · Analog Circuits and Signal Processing,
https://doi.org/10.1007/978-3-030-95284-6_1

tags, which usually do not have batteries [11, 12], whereas a large temperature range is required in automotive and industrial ICs [13]. In general, high accuracy is desirable in most applications and must be accompanied by commensurate resolution to facilitate practical calibration.

Resolution is defined as the minimum temperature change that can be detected by a sensor, and it is typically limited by random noise [4]. In many applications, high resolution is not a critical requirement, 1 °C resolution, for example, is more than sufficient for cooking ovens and coffee machines. However, industrial applications, for example, the temperature control of wafer steppers, often require much higher resolution. Because the position of wafer steppers must be controlled with nanometer precision, the thermal expansion of their mechanical components should be carefully controlled and minimized. The required temperature sensing accuracy is then at the mK level, while the sensing resolution should then be at the sub-mK level [14]. As will be discussed in the next section, similar levels of resolution are required for the temperature compensation of frequency references.

1.2 Challenges in Frequency Reference Compensation

The performance of electronic systems often relies on the accuracy and noise of clock references. For instance, the USB 3.0 serial bus standard requires clocks with less than 50 ppm frequency error, and less than 0.8 ps (~16 ppm) jitter [15]. For telecommunication systems, the requirements are even stricter: less than 0.1 ppm frequency error and an Allan Deviation (a measure of long-term stability) below 10^{-10} in an integration time of 1 s [5].

Typically, accurate clock references are based on quartz crystal oscillators. Recently, references based on MEMS (Micro Electro Mechanical System) [5, 6] or BAW (bulk acoustic wave) [16] resonators have become popular due to their small size and ease of integration with CMOS technology. However, their resonant frequencies are significantly temperature dependent. For example, the temperature coefficient (TC) of an uncompensated MEMS oscillator is about 31 ppm/°C [6], resulting in a 4000 ppm frequency change from −40 °C to 85 °C. Even for quartz crystal oscillators [7], or BAW devices with passive compensation schemes [16], errors of ~200 ppm may occur over the same temperature range. As a consequence, temperature compensation must be included to achieve high frequency accuracy, especially for MEMS oscillators.

Figure 1.1 shows the block diagram of a typical high-accuracy MEMS-based frequency reference [6]. It contains a MEMS oscillator, a fractional-N synthesizer and divider, a compensating temperature sensor, and digital processing blocks. The output of the temperature sensor is used to control the output frequency of the fractional-N synthesizer via a polynomial engine, in order to compensate for the temperature dependence of the resonator. This temperature dependence is typically significantly nonlinear, and spreads over samples, thus necessitating multiple-point calibration. In [6], a resistor-based temperature sensor and a 5th-order polynomial

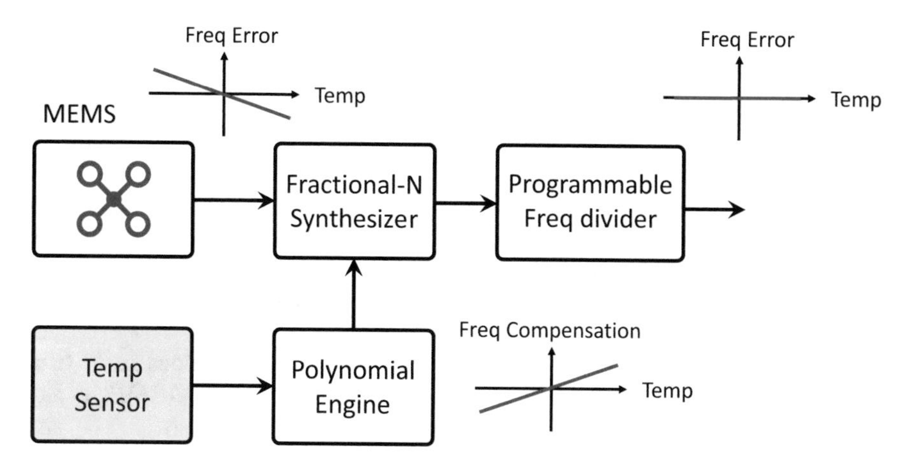

Fig. 1.1 Temperature compensation of a MEMS-based frequency reference

are used to achieve less than 0.5 ppm frequency error from −40 °C to 85 °C. In a more recent design [5], a MEMS-based temperature sensor and a 7th-order polynomial are used to reduce frequency error to the extraordinary level of 40 ppb.

Since the temperature sensor's noise is injected into the fractional-N synthesizer, it appears as phase noise in the output frequency. To minimize this, a high-resolution temperature sensor is required. For example, in the 40 ppb inaccuracy MEMS frequency reference reported in [5], better than 100 μK resolution was required. In a less accurate design, with 0.3 ppm frequency inaccuracy, the required resolution is still quite high: better than 650 μK [8].

Apart from high resolution, temperature sensors intended for frequency references should also achieve a bandwidth of about 100 Hz to maintain frequency accuracy in the presence of temperature variations [5]. Moreover, a low-power sensor is preferred to minimize its contribution to the total energy budget, meaning that its energy efficiency should be high. This also helps to reduce self-heating, which may lead to temperature compensation errors.

To reduce their overall manufacturing cost, the multiple-point calibration of temperature sensors is generally not desirable. However, since MEMS oscillators typically require multiple-point calibration anyway, the extra cost of temperature sensor calibration becomes almost negligible. The achievable accuracy of such sensors, and thus of the resulting frequency references, will then be mainly limited by their long-term drift.

1.3 Resolution and Resolution FoM

Of the specifications introduced above, energy efficiency remains a vague term without a clear metric. To quantify it, a resolution figure of merit (FoM) [4] has been defined in the same way as the Schreier FoM for ADCs [17]. For temperature sensors, this can be expressed as:

$$\mathrm{FoM} = \mathrm{Energy} / \mathrm{Conversion} \cdot \mathrm{Resolution}^2 \tag{1.1}$$

In this equation, the resolution is squared to reflect the fact that the resolution of a temperature sensor should be limited by thermal noise. Therefore, to obtain 2× more resolution, the energy consumption of a sensor should be increased by 4×. With this metric, higher energy efficiency corresponds to a smaller resolution FoM. Compared to resolution, the resolution FoM of a temperature sensor is a more fundamental specification, because as long as the sensor is thermal-noise limited, better resolution can always be obtained at the expense of increased energy consumption.

Figure 1.2 shows the resolution and the resolution FoM of various smart temperature sensors at the start of the work described in this work (in 2016). The plot includes sensors based on BJTs, MOSFETs, resistors, thermal diffusivity (TD), and MEMS devices. Although MEMS-based sensors [19] can achieve superb resolution (40 μK) and energy efficiency (120 fJ·K^2), their non-CMOS fabrication leads to two-die systems, greater complexity, and increased cost. Of the possible CMOS-compatible candidates, BJT- and resistor-based sensors achieve mK-level resolution and good energy efficiency (~1 pJ·K^2), while MOSFET-based sensors do not have sufficient resolution and TD sensors have relatively poor efficiency.

To understand the limitations of the different types of sensors and identify the most suitable one for the targeted frequency compensation application, the operating principles of different types of sensors will be discussed in the next section, as well as the theoretical limits on their energy efficiency.

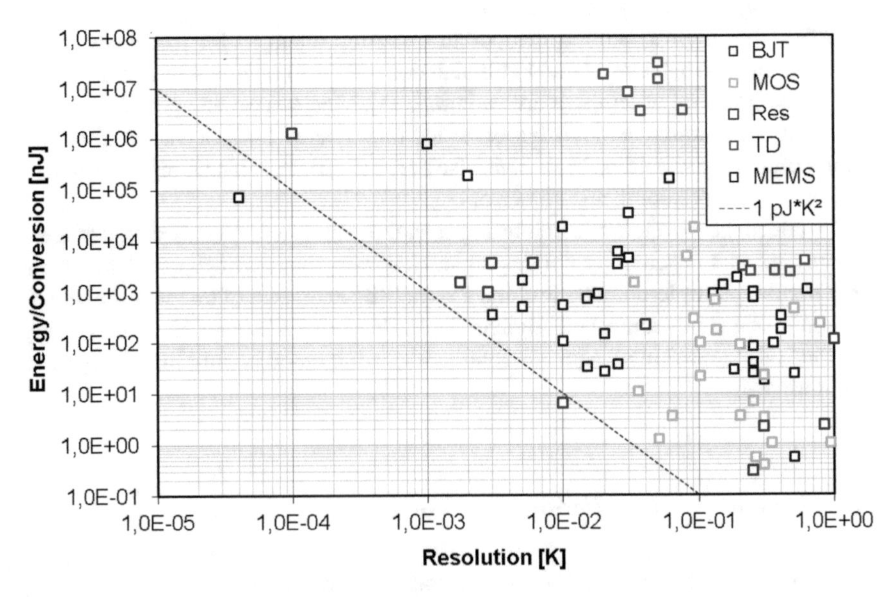

Fig. 1.2 Energy per conversion vs. resolution of temperature sensors, published prior to the start of this research (in 2016) [18]

1.4 CMOS Temperature Sensing Elements and Their Theoretical Resolution FoMs

1.4.1 Bipolar Junction Transistors (BJTs)

Because of a lower sensitivity to process spread and packaging stress, vertical BJTs are preferred over lateral ones in temperature sensors [20]. Both PNPs [21] and NPNs [22] can be used to sense temperature. The NPN transistor is the more ideal candidate because of its larger current gain. However, it requires a deep N-well option, which is not always available in modern CMOS processes (Fig. 1.3).

Regardless of the type of BJT used, its base-to-emitter voltage V_{BE} can be approximated over a wide range of collector currents as:

$$V_{BE} = \frac{kT}{q} \cdot \ln\left(\frac{I_C}{I_S}\right), \tag{1.2}$$

where I_C (>>I_S) is the collector current, I_S is the saturation current, k is the Boltzmann constant, q is the electron charge, and T is the absolute temperature.

For a pair of BJTs biased with different current densities, the V_{BE} difference can be expressed as:

$$\Delta V_{BE} = \frac{kT}{q} \cdot \ln(p \cdot r), \tag{1.3}$$

where p and r are the ratios of the collector current and emitter area between two bipolar transistors, respectively. This is known as a PTAT (proportional-to-absolute-temperature) voltage.

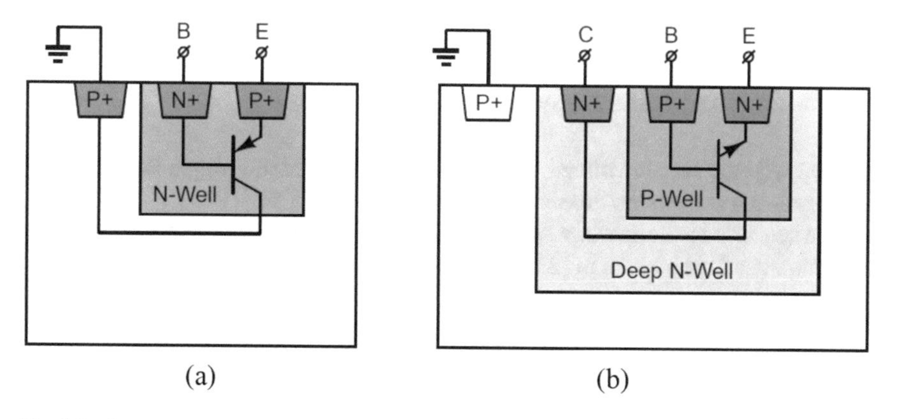

Fig. 1.3 Cross section of (**a**) a vertical PNP transistor in standard CMOS and (**b**) a vertical NPN transistor in CMOS technology with a deep N-well option

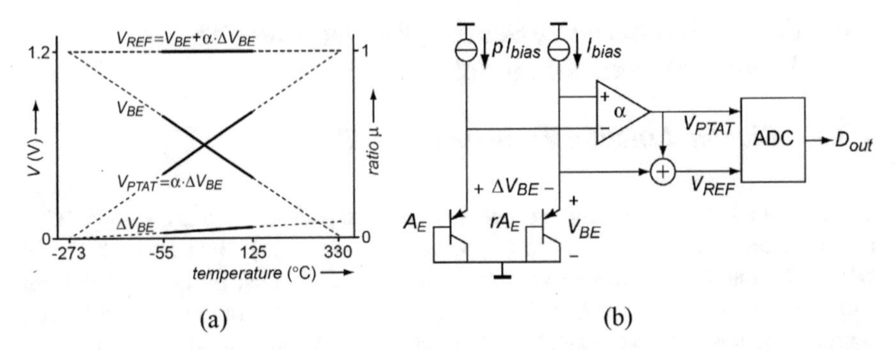

Fig. 1.4 (**a**) V_{BE}, ΔV_{BE} and V_{Ref} over temperature; (**b**) simplified block diagram of a BJT-based temperature sensor [21]

A reference voltage is needed to digitize ΔV_{BE}. This is usually achieved by linearly combining V_{BE} and ΔV_{BE}. The ratio of ΔV_{BE} and the well-known bandgap voltage V_{REF} ($\approx$1.22 V) is then a function of absolute temperature given by:

$$\mu = \frac{V_{\text{PTAT}}}{V_{\text{REF}}} = \frac{\alpha \Delta V_{\text{BE}}}{V_{\text{BE}} + \alpha \Delta V_{\text{BE}}}, \tag{1.4}$$

where α is the ΔV_{BE} scaling factor inversely proportional to $\ln(p\cdot r)$. Using PNP transistors as an example, the relationship between the voltages mentioned above and the circuit used to provide μ is shown in Fig. 1.4.

To achieve high accuracy, dynamic element matching is often used to cancel the mismatch between the BJTs and the current sources [21]. To avoid complex logic, the ratios (p and r) should be kept small. In most BJT sensors, $p\cdot r$ is between 2 and 10. With $p = 5$, for example, the sensitivity of ΔV_{BE} is 0.14 mV/K.

The theoretical energy efficiency of BJT sensors is limited by its front-end. As a first step, we assume that ΔV_{BE} is realized by scaling the current mirror ratio p, and that the readout circuit contributes zero power and noise. The sensor's energy consumption can then be expressed as:

$$E_{\text{conv}} = V_{\text{DD}} \cdot (1+p) \cdot I_{\text{bias}} \cdot T_{\text{conv}}, \tag{1.5}$$

where V_{DD} is the supply voltage, I_{bias} is the biasing current of a single BJT, and T_{conv} is the sensor's conversion time.

The sensor's resolution is mainly limited by the noise present in ΔV_{BE}, as this is much smaller than V_{BE}. As in [23], this noise is given by:

$$v^2_{\text{n},\Delta V_{\text{BE}}} = \frac{4kT}{g_m} B_{\text{n}} \cdot \left(1 + \frac{1}{p}\right), \tag{1.6}$$

where $B_{\text{n}} = 1/(2\cdot T_{\text{conv}})$ is the noise bandwidth.

After computing the sensor's resolution, using its sensitivity and the calculated noise, and combining it with the sensor's energy consumption, both T_{conv} and I_{bias} will cancel out, resulting in the following expression for the sensor's FoM [23]:

$$\mathrm{FoM}_{\mathrm{BJT,p}} = 2 \cdot \frac{(p+1)^2}{p} \cdot V_{\mathrm{DD}} \cdot q \cdot V_{\mathrm{T}}^2 \cdot \alpha^2 \cdot \left(\frac{A-T}{V_{\mathrm{REF}}}\right)^2, \tag{1.7}$$

where $V_{\mathrm{T}} = kT/q$ and $A \approx 2\mathrm{T} = 600$ K. Assuming $p = 5$ and a supply voltage of $V_{\mathrm{DD}} = 1.8$ V, the theoretical FoM of this BJT sensor configuration is approximately 36 fJ·K^2.

Alternatively, one can fix $p = 1$ and scale the emitter area ratio r. Due to the more balanced power and noise distribution of the two BJTs, the theoretical energy efficiency improves. According to [23], this can be calculated as:

$$\mathrm{FoM}_{\mathrm{BJT,r}} = 8 \cdot V_{\mathrm{DD}} \cdot q \cdot V_{\mathrm{T}}^2 \cdot \alpha^2 \cdot \left(\frac{A-T}{V_{\mathrm{REF}}}\right)^2. \tag{1.8}$$

With $r = 5$ and the same $V_{\mathrm{DD}} = 1.8$ V, the theoretical FoM is reduced to 20 fJ·K^2.

Note that in the computation of these FoMs, the power and noise of the readout circuit have been neglected. Assuming that both the readout circuit and the BJT sensor front-end have similar power/noise levels, the FoM of a complete BJT-based temperature sensor will then be 4× larger, that is, 80 fJ·K^2. Back in 2016, the state-of-the-art FoM achieved by BJT-based sensors was 3.6 pJ·K^2 [24].

1.4.2 MOSFETs

The behavior of a MOSFET device (transconductance, threshold voltage, etc.) heavily depends on temperature, and hence there are various types of MOS-based temperature sensors. They can be roughly grouped as:

1. BJT-like sensors
2. Other subthreshold region sensors
3. Saturation region sensors

In BJT-like designs, which achieve by far the best resolution and efficiency performance, two MOSFETs, biased in the subthreshold (weak-inversion) region, serve as replacements of BJT devices in Fig. 1.3 [25]. In particular, the sensor's inaccuracy can be improved by configuring the MOSFETs as dynamic-threshold MOSTs (DTMOSTs) [25, 26], in which the transistor's body is connected to its gate to provide a well-defined threshold voltage, as shown in Fig. 1.5. In both cases, the *V-I* characteristic is similar compared to that of a BJT device, that is:

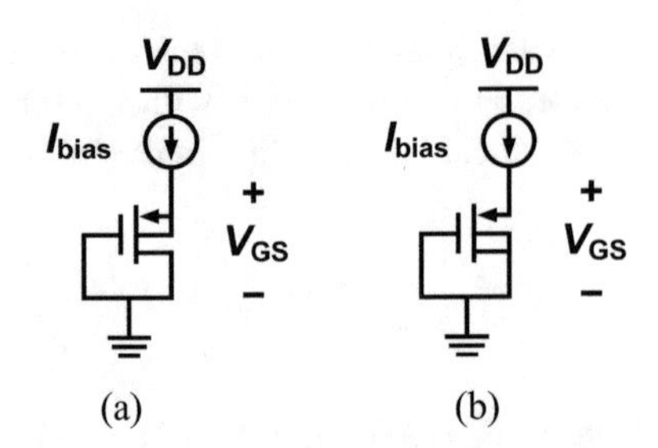

Fig. 1.5 A PMOS transistor configured as (**a**) a diode and (**b**) a DTMOST

$$I_{DS} = I_{D0} \cdot e^{\frac{V_{GS}-V_T}{nkT/q}}$$
$$V_{GS} = \frac{nkT}{q} \cdot \ln\left(\frac{I_{DS}}{I_{D0}}\right) + V_T, \tag{1.9}$$

where V_{GS} is the gate-source voltage, V_T is the threshold voltage, n is the slope factor, and I_{D0} is the current at $V_{GS} = V_T$. Therefore, the analysis in Sect. 1.4.1 still holds.

Compared to BJT-based sensors, two variables change in the theoretical FoM calculation: the supply voltage and the noise. Since V_{GS} ($\approx$0.4 V) is typically smaller than V_{BE} ($\approx$0.7 V), MOSFET-based sensors can work with a smaller supply voltage, potentially below 1 V. However, given the same biasing current, the gm of a subthreshold MOSFET is n times smaller than that of a BJT, and the thermal noise indicated by Eq. (1.6) is proportionally worse. Also, MOSFETs typically exhibit much more 1/f noise. Due to these factors, the theoretical FoM of such MOSFET-based sensors is somewhat worse than that of BJT-based designs. However, back in 2016, the state-of-the-art FoM for MOSFET sensors was 3.2 pJ·K^2 [27].

However, the resolution of these sensors is typically not sufficiently high for frequency compensation applications. This is mainly due to the power restriction posed by the weak-inversion operation: with the same theoretical FoM and conversion speed, the resolution of these low-power (typically <1 μW) MOSFET sensors is definitely lower than their BJT-based counterparts, the exponential V-I characteristic of which can be maintained at a mW power level [28].

Some subthreshold MOSFET sensors are based on other temperature sensing principles. For example, [29] utilizes the temperature sensitivity of V_T. This linear sensitivity, denoted as κ_{VT} (i.e., $V_T(T) = V_{T0} + \kappa_{VT} \cdot \Delta T$), results in an exponential variation on I_{DS} with temperature. As shown in Fig. 1.6, this current is further proportionally converted to frequency using a ring oscillator, and, with the help of a reference frequency, is digitized by a counter-based readout circuit.

To accurately determine the theoretical FoM of such sensors, the temperature dependency of I_{D0} should be also taken into account, which greatly complicates the temperature sensitivity expression of I_{DS}. For simplicity, this sensitivity κ_{IDS} is typically derived from simulations, and is roughly 4.6%/°C at room temperature [30].

Neglecting the power of the readout circuit, the energy consumption of such sensors can be simplified as:

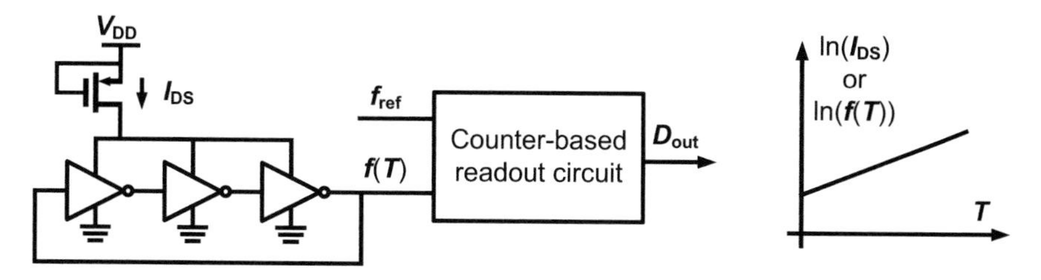

Fig. 1.6 Simplified diagram of a subthreshold MOSFET temperature sensor with frequency-based readout [29]

$$E_{\mathrm{conv}} = V_{\mathrm{DD}} \cdot I_{\mathrm{DS}} \cdot T_{\mathrm{conv}}. \tag{1.10}$$

And the current noise of I_{DS} is given by:

$$i_{\mathrm{n}}^2 = 4kTg_{\mathrm{m}} \cdot B_{\mathrm{n}} = \frac{2q \cdot I_{\mathrm{DS}}}{n \cdot T_{\mathrm{conv}}}. \tag{1.11}$$

After neglecting the area-dependent $1/f$ noise and assuming this thermal noise to be the only noise source, the theoretical FoM of the MOSFET sensor front-end can be calculated as:

$$\mathrm{FoM}_{\mathrm{MOS}} = \frac{q \cdot V_{\mathrm{DD}}}{2n \cdot \kappa_{\mathrm{IDS}}^2} \tag{1.12}$$

Assuming $n = 1.2$, and $V_{\mathrm{DD}} = 1$ V, the theoretical FoM is roughly 0.03 fJ·K², which is almost 1000× better than that of the BJT-like sensors.

However, these sensors are not well suited for the temperature compensation of frequency references. First, the oscillator-based readout circuit typically introduces significant excess power and noise, which severely degrades the energy efficiency. In a relatively efficient design based on a native NMOS current source, the FoM turns out to be 3200 fJ·K² [31]. Second, like BJT-like sensors, the sensor's power, and thus its resolution within a certain conversion time, is limited by the subthreshold operation. Last but not the least, sensor performance, including power, resolution, and sensitivity, varies exponentially over temperature. This is not conducive to realizing frequency references with stable performance over temperature.

The theoretical resolution of MOSFET sensors working in saturation regions is not restricted by the aforementioned power problem. However, existing designs are still not suitable for the targeted frequency compensation application. These sensors are mostly based on compact ring oscillators and counter-based readout, resulting in a small chip area [32, 33]. In [32], for instance, the temperature is obtained by digitizing the frequency ratio $f_1(T) / f_2(T)$ of two ring oscillators, as shown in Fig. 1.7. The two ring oscillators are comprised of transistors with different threshold voltages, leading to different effects on mobility variation and thus different

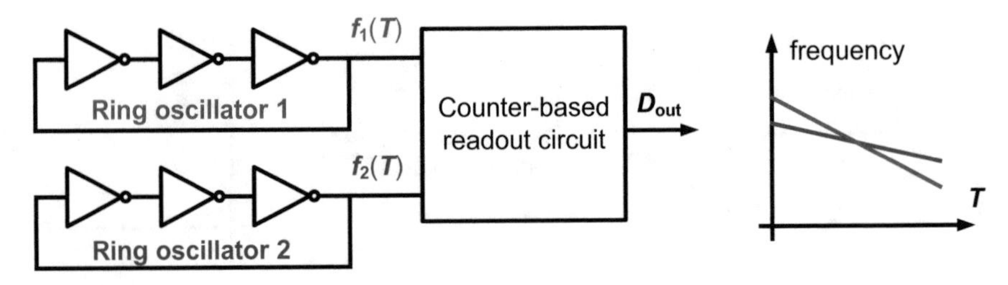

Fig. 1.7 Simplified diagram of the MOSFET temperature sensor in [32]

temperature sensitivities. The temperature is then digitized by calculating the oscillator frequency ratio using counters. Unfortunately, these sensors are extremely vulnerable to supply variations (1–10 °C/V). Also, their low resolution (>100 mK) is usually limited by the counter's quantization noise instead of the thermal noise of their front-end, so that their FoM is far worse than the theoretical limit.

Due to these drawbacks, MOSFET-based temperature sensors are mainly used in system-on-a-chip (SoC) applications with limited supply voltages. Also, they typically require less chip area than BJTs, which helps to reduce the fabrication cost.

1.4.3 Electro-thermal Filters (ETFs)

Temperature sensors based on electro-thermal filters make use of the well-defined and temperature-dependent speed at which heat diffuses through a silicon substrate [34–36]. The thermal diffusivity of silicon, denoted as D_{si}, can be approximated by the power law of $D_{si} \propto 1/T^{1.8}$ [37].

The structure of an electro-thermal filter is shown in Fig. 1.8. It consists of a heater that generates heat pulses, and a relative temperature sensor (thermopile) located at a distance s from the heater which detects the heat propagation delay in between these two elements. The sensor has a square-wave voltage input, and the generated heat pulses are converted back to a small voltage signal by the thermopile. As a result, the ETF behaves like a low-pass filter in the time domain (Fig. 1.9). Given a fixed excitation frequency f_{drive}, its phase shift can be expressed as:

$$\varnothing_{ETF} = -s\sqrt{\pi f_{drive} / D_{si}} \propto -s\sqrt{\pi f_{drive} T^{1.8}} \tag{1.13}$$

A phase-domain ADC can then be used to digitize this phase and obtain the temperature information.

Because on-chip thermopiles usually have low sensitivity (~0.5 mV/K), the output level of ETFs is rather small (~1 mV), which, in turn, severely limits their resolution. Together with the power dissipation of their heaters (several mWs), the result is poor energy efficiency. At the start of this research (2016), the best reported FoM for an ETF was $1.4 \cdot 10^5$ pJ·K^2 [36], which is about 10^5 times worse than the record FoMs obtained for BJT- and resistor-based temperature sensors at that time.

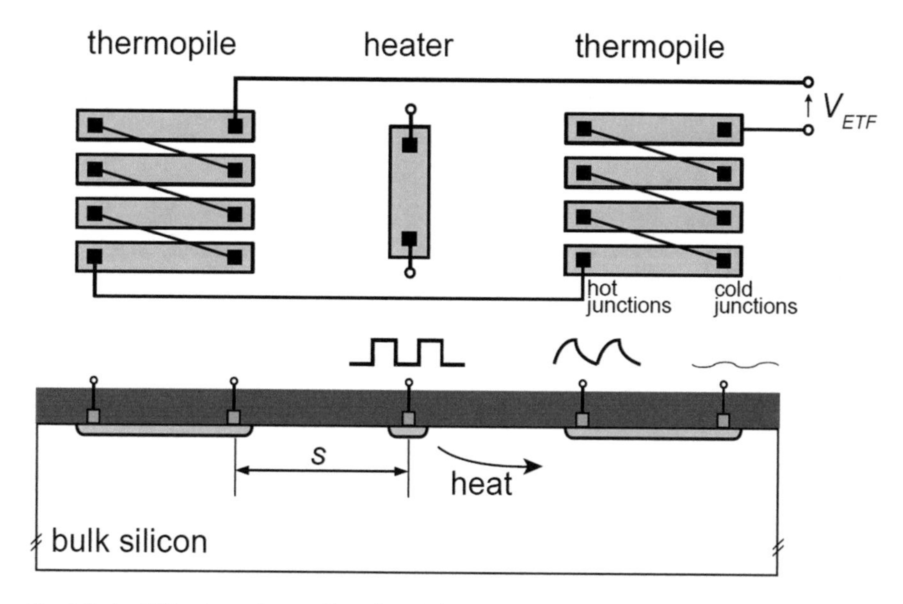

Fig. 1.8 An ETF using a thermopile as its relative temperature sensor [35]

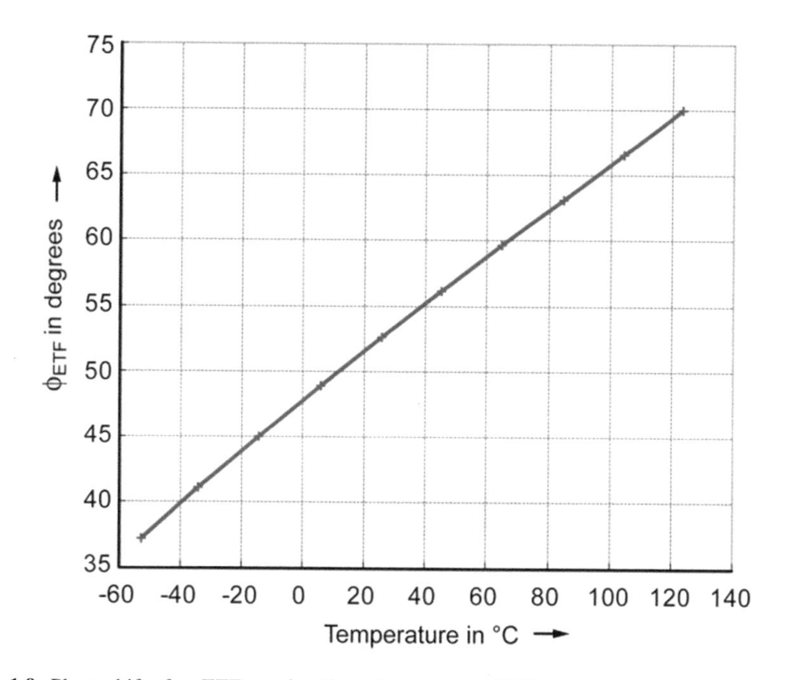

Fig. 1.9 Phase shift of an ETF as a function of temperature [35]

Although ETF-based temperature sensors are highly energy inefficient, they can achieve both high accuracy and small chip area [36] due to the well-defined D_{si}. This makes such sensors promising in dense thermal management applications, where high power and low resolution can be tolerated.

1.4.4 Resistors

Most CMOS-compatible resistors exhibit significant temperature coefficients (TCs). According to [39], the temperature dependence of a resistor can be well modeled as:

$$R_S(T) = R_S(T_0) \cdot \left(1 + \mathrm{TC}_{S1} \cdot \Delta T + \mathrm{TC}_{S2} \cdot \Delta T^2\right) \tag{1.14}$$

where $R_S(T_0)$ is the nominal resistance at a reference temperature T_0, TC_{S1} and TC_{S2} are its 1st- and 2nd-order TCs, and ΔT is the temperature with respect to T_0. Figure 1.10 shows the temperature characteristics of different resistors in a standard 0.18 μm CMOS technology. Depending on the resistor type, the 1st-order TC at room temperature (~25 °C) ranges from −0.15%/°C to 0.34%/°C.

Given a certain resistor type, $R_S(T_0)$, TC_{S1} and TC_{S2} all spread, which necessitates a multi-point calibration [6, 10]. Among those variables, $R_S(T_0)$ spreads the most

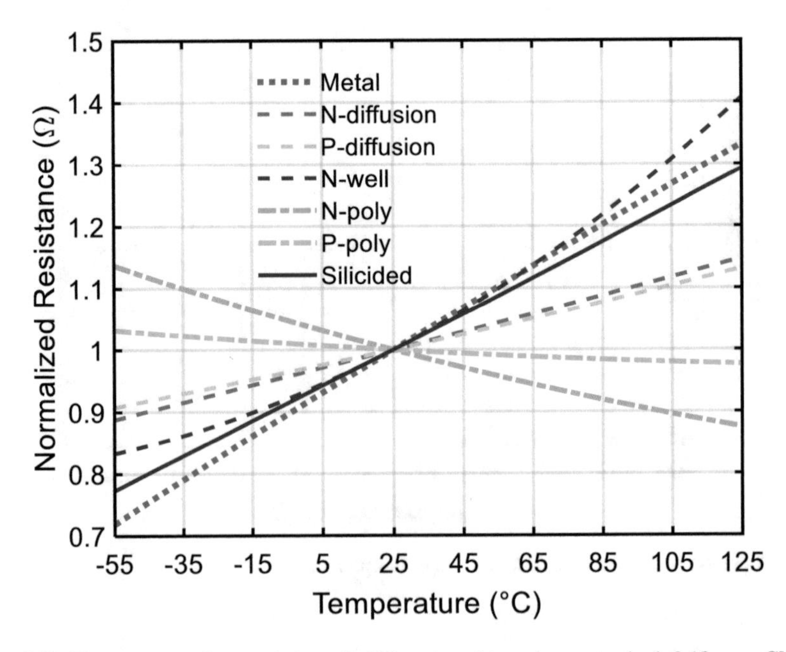

Fig. 1.10 Temperature characteristics of different resistors in a standard 0.18 μm CMOS technology

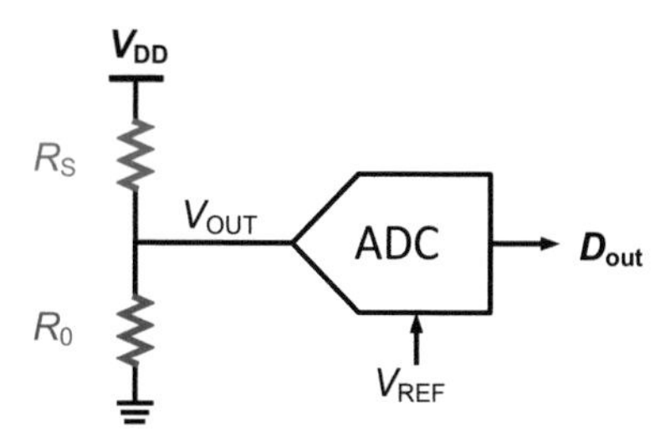

Fig. 1.11 Temperature sensor based on a resistor divider

(~±20% over corners), TC_{S1} spreads less, while the spread of TC_{S2} is often negligible. As a result, a well-designed sensor should achieve good accuracy with only 2 trimming points.

Figure 1.11 shows a simple temperature sensor that consists of a temperature-sensitive resistor (thermistor) R_S and a reference resistor R_0. The temperature-dependent voltage output of the resistor divider is then digitized by an ADC. Here it is assumed that R_0 has a zero TC, and the temperature dependence of R_S is given by Eq. (1.14). With a small temperature difference of ΔT, the voltage output under a supply voltage of V_{DD} can then be expressed as:

$$V_{OUT} = V_{DD} \cdot \frac{1}{2 + \Delta T \cdot TC_{S1}}. \tag{1.15}$$

With $V_{DD} = 1.8$ V and $TC_{S1} = 0.3\%/°C$, the V_{OUT} sensitivity is 1.35 mV/°C. In the case of a balanced bridge (i.e., $\Delta T = 0$), the resolution FoM of the sensor front-end can be calculated as [38]:

$$FoM_{RES} = \frac{8kT}{TC_{S1}^{\ 2}}. \tag{1.16}$$

With $TC_{S1} = 0.3\%/°C$, the theoretical FoM is about 3.7 fJ·K^2. Considering the 4× factor from the ADC's power and noise, the practical FoM limit becomes 15 fJ·K^2, which is 5× better compared to that of BJT-based sensors. Note that, unlike BJT-based sensors, the theoretical FoM of a resistor-based sensor only depends on the thermistor's TC, but it is independent of its supply voltage or biasing current. At the start of this research (2016), the best reported FoM for a resistor-based sensor was 0.65 pJ·K^2 [39].

Other than the high-resolution FoM, resistor-based sensors can be easily scaled along constant-FoM lines to achieve either high-resolution or low-power requirements. Also, there is no minimum supply limitation. A drawback of these sensors, as mentioned before, is the need for multi-point calibration.

1.5 Choice of the Sensing Element

After reviewing the principles and limitations of different CMOS temperature sensing elements, resistor-based sensors are the most suitable choice for the targeted application of compensating the temperature dependency of frequency references. Their practical energy efficiency is the best among all CMOS candidates. Although the need for multi-point calibration is a disadvantage, it can be tolerated as such calibration is needed anyway for the oscillator.

It would be interesting to investigate energy-efficient sensors built with other sensing elements, such as BJTs or MOSFETs working in saturation regions. However, this is beyond the scope of this book.

1.6 Goals and Book Organization

The main goal of this book is to provide a comprehensive study of different types of resistor-based temperature sensors, and to compare their pros and cons with respect to conventional BJT-based temperature sensors. In Chap. 2, the critical choices involved with their design will be discussed, together with a literature review.

Another goal is the development of resistor-based temperature sensors that can be used for the temperature compensation of frequency references. Among all the structures, two major types—namely Wien bridge (WB) sensors and Wheatstone bridge (WhB) sensors—have been designed, fabricated, and characterized. The Wien bridge sensors, which achieve better accuracy but worse energy efficiency, are presented in Chap. 3. The Wheatstone bridge designs, which use various design techniques to improve the sensor's resolution FoM, are presented in Chap. 4. Ultimately, a 10 $fJ{\cdot}K^2$ FoM has been achieved, which improves the state-of-the-art by 65×.

Last but not least, two resistor-based temperature sensors designed to broaden the application of resistor-based sensors are presented in Chap. 5. One is a low-power sensor designed for biomedical applications, and the other is a sensor integrated into an RC-based frequency reference.

After discussing how these results were achieved, this book ends with Chap. 6, which consists of conclusions and discussions of future work.

References

1. "NTCLG100E2 datasheet", Vishay, Inc., July 2015, www.vishay.com
2. "B59100 datasheet", EPCOS AG, Nov. 2013., www.tdk-electronics.tdk.com
3. "KA01 datasheet", T. M. Electronics ltd., www.tmelectronics.com
4. K.A.A. Makinwa, "Smart temperature sensors in standard CMOS," (Proc. Eurosensors). Proc. Eng **5**, 930–939 (2010)

5. M.H. Roshan et al., A MEMS-assisted temperature sensor with 20-μK resolution, conversion rate of 200 S/s, and FOM of 0.04 pJK2. IEEE J. Solid State Circuits **52**(1), 185–197 (2017)
6. M.H. Perrott et al., A temperature-to-digital converter for a MEMS-based programmable oscillator with <±0.5-ppm frequency stability and <1-ps integrated jitter. IEEE J. Solid State Circuits **48**(1), 276–291 (2013)
7. D. Ruffieux et al., A 3.2 × 1.5 × 0.8 mm3 240 nA 1.25-to-5.5V 32 kHz-DTCXO RTC module with an overall accuracy of ±1 ppm and an all-digital 0.1 pm compensation-resolution scheme at 1 Hz, in *IEEE ISSCC Dig. Tech. Papers*, (2016, Jan), pp. 208–209
8. S.H. Shalmany et al., A 620μW BJT-based temperature-to-digital converter with 0.65mK resolution and FoM of 190fJ·K^2, in *IEEE ISSCC Dig. Tech. Papers*, (2020, Feb), pp. 70–71
9. Z. Wang et al., An in-situ temperature-sensing interface based on a SAR ADC in 45nm LP digital CMOS for the frequency-temperature compensation of crystal oscillators, in *IEEE ISSCC Dig. Tech. Papers*, (2010, Feb), pp. 316–317
10. P. Park, D. Ruffieux, K.A.A. Makinwa, A thermistor-based temperature sensor for a real-time clock with ±2 ppm frequency stability. IEEE J. Solid State Circuits **50**(7), 1571–1580 (2015)
11. M.K. Law, A. Bermak, H.C. Luong, A sub-μW embedded CMOS temperature sensor for RFID food monitoring application. IEEE J. Solid State Circuits **45**(6), 1246–1255 (2010)
12. J. Yin et al., A system-on-chip EPC Gen-2 passive UHF RFID tag with embedded temperature sensor, in *IEEE ISSCC Dig. Tech. Papers*, (2010, Feb), pp. 308–309
13. K. Souri, K. Souri, K. Makinwa, A 40μW CMOS temperature sensor with an inaccuracy of ±0.4°C (3σ) from −55°C to 200°C, in *IEEE Proc. ESSCIRC*, (2013, Sept), pp. 221–224
14. B. Parekh et al., UPW immersion lithography: Purification needs and solutions, in *Ultrapure Fluid Handling and Wafer Cleaning Conference*, (Feb. 2008)
15. "TUSB 1310 data manual", Texas Instruments Inc., May 2011, www.ti.com
16. D. Griffith et al., An integrated BAW oscillator with <±30ppm frequency stability over temperature, package stress, and aging suitable for high-volume production, in *IEEE ISSCC Dig. Tech. Papers*, (2020, Feb), pp. 58–60
17. R. Schreier, G.C. Temes, *Understanding Delta-Sigma Data Converters* (Wiley, New York, 2005)
18. K.A.A. Makinwa, "Smart temperature sensor survey", [Online]. Available: http://ei.ewi.tudelft.nl/docs/TSensor_survey.xls
19. M.H. Roshan et al., Dual-MEMS-resonator temperature-to-digital converter with 40μK resolution and FOM of 0.12pJK^2, in *IEEE ISSCC Dig. Tech. Papers*, (2016, Feb), pp. 200–201
20. J.F. Creemer, F. Fruett, G.C.M. Meijer, P.J. French, The piezojunction effect in silicon sensors and circuits and its relation to piezoresistance. IEEE Sensors J. **1**(2), 98–108 (2001)
21. M.A.P. Pertijs, K.A.A. Makinwa, J.H. Huijsing, A CMOS smart temperature sensor with a 3σ inaccuracy of ±0.1°C from −55°C to 125°C. IEEE J. Solid State Circuits **40**(12), 2805–2815 (2005)
22. F. Sebastiano, L.J. Breems, K.A.A. Makinwa, S. Drago, D.M.W. Leenaerts, B. Nauta, A 1.2-V 10μW NPN-based temperature sensor in 65-nm CMOS with an inaccuracy of 0.2°C (3σ) from −70 °C to 125 °C. IEEE J. Solid State Circuits **45**(12), 2591–2601 (2010)
23. K. Souri, K.A.A. Makinwa, Readout methods for BJT-based temperature sensors, in *Energy-Efficient Smart Temperature Sensors in CMOS Technology (Analog Circuits and Signal Processing)*, (Springer, Cham, 2018)
24. A. Heidary, G. Wang, K. Makinwa, G. Meijer, A BJT-based CMOS temperature sensor with a 3.6pJ·K^2-resolution FoM, in *IEEE ISSCC Dig. Tech. Papers*, (2014, Feb), pp. 224–225
25. M. Terauchi, Selectable logarithmic/linear response active pixel sensor cell with reduced fixed-pattern-noise based on dynamic threshold MOS operation. Jpn. J. Appl. Phys. **44**(4B), 2347–2350 (2005)
26. K. Souri, Y. Chae, F. Thus, K. Makinwa, A 0.85V 600nW all-CMOS temperature sensor with an inaccuracy of ±0.4°C (3σ) from −40 to 125°C, in *IEEE ISSCC Dig. Tech. Papers*, (2014, Feb), pp. 222–223
27. Y. Kim et al., A 0.02mm^2 embedded temperature sensor with ±2°C inaccuracy for self-refresh control in 25nm mobile DRAM, in *IEEE ESSCIRC*, (2015, Sept), pp. 267–270
28. J. Shor, K. Luria, D. Zilberman, Ratiometric BJT-based thermal sensor in 32nm and 22nm technologies, in *IEEE ISSCC Dig. Tech. Papers*, (2012, Feb), pp. 210–212

29. E. Saneyoshi, K. Nose, M. Kajita, M. Mizuno, A 1.1V 35μm × 35μm thermal sensor with supply voltage sensitivity of 2°C/10%-supply for thermal management on the SX-9 supercomputer, in *IEEE Symp. VLSI Circ*, (2008, June), pp. 152–153
30. Z. Tang, Y. Fang, Z. Shi, X. Yu, N.N. Tan, W. Pan, A 1770- μm^2 leakage-based digital temperature sensor with supply sensitivity suppression in 55-nm CMOS. IEEE J. Solid State Circuits **55**(3), 781–793 (2020)
31. K. Yang et al., A 0.6nJ −0.22/+0.19°C inaccuracy temperature sensor using exponential subthreshold oscillation dependence, in *IEEE ISSCC Dig. Tech. Papers*, (2017, Feb), pp. 160–161
32. T. Anand, K.A.A. Makinwa, P.K. Hanumolu, A VCO based highly digital temperature sensor with 0.034 °C/mV supply sensitivity. IEEE J. Solid State Circuits **51**(11), 2651–2663 (2016)
33. D. Ha, K. Woo, S. Meninger, T. Xanthopoulos, E. Crain, D. Ham, Time-comain CMOS temperature sensors with dual delay-locked loops for microprocessor thermal monitoring. IEEE Trans. VLSI Syst **20**(9), 1590–1601 (2012)
34. K.A.A. Makinwa, M.F. Snoeij, A CMOS temperature-to-frequency converter with an inaccuracy of less than ±0.5°C (3σ) from −40°C to 105°C. IEEE J. Solid State Circuits **41**(12), 2992–2997 (2006)
35. C.P.L. van Vroonhoven, D. d'Aquino, K.A.A. Makinwa, A thermal-diffusivity-based temperature sensor with an untrimmed inaccuracy of ±0.2°C (3σ) from −55°C to 125°C, in *IEEE ISSCC Dig. Tech. Papers*, (2010, Feb), pp. 314–315
36. U. Sönmez, F. Sebastiano, K.A.A. Makinwa, Compact thermal-diffusivity-based temperature sensors in 40-nm CMOS for SoC thermal monitoring. IEEE J. Solid State Circuits **52**(3), 834–843 (2017)
37. T. Veijola, Simple model for thermal spreading impedance, in *Proc. BEC*, (1996), pp. 73–76
38. S. Pan, K.A.A. Makinwa, Energy-efficient high-resolution resistor-based temperature sensors, in *Hybrid ADCs, Smart Sensors for the IoT, and Sub-1V & Advanced Node Analog Circuit Design*, (Springer, 2018), pp. 183–200
39. C.H. Weng, C.K. Wu, T.H. Lin, A CMOS thermistor-embedded continuous-time delta-sigma temperature sensor with a resolution FoM of 0.65 pJ °C². IEEE J. Solid State Circuits **50**(11), 2491–2500 (2015)

Chapter 2
Sensor and Readout Topologies

2.1 Introduction

This chapter discusses some general issues involved in the design of resistor-based temperature sensors. First, the characteristics of the different sensing resistors available in standard CMOS technology are described. This is followed by a discussion of the available impedance references, which are needed to convert resistance changes into digital information. Two possible sensor structures: dual-R (with a resistor reference) and RC (with a capacitor reference) are then presented. Finally, the requirements and architectures of various readout circuits are discussed.

2.2 Sensor Design

2.2.1 Sensing Resistors

In standard CMOS technology, many types of resistors are available, including metal resistors, diffusion resistors, polysilicon (poly) resistors, N-well resistors, and silicided resistors. As shown in Fig. 1.11, they all are temperature dependent to some degree and so can all be potentially used to sense temperature. Although resistors can be also realized with active devices, for example, by biasing a MOSFET in the triode region, they suffer from a greater spread, and are much less stable.

To meet the stringent resolution, energy efficiency, and stability requirements of the targeted application—the temperature compensation of frequency references, sensing resistors with a large temperature coefficient (TC), low $1/f$ noise, and high stability (low voltage and stress sensitivities, low long-term drift) should be used.

S. Pan, K. A. A. Makinwa, *Resistor-based Temperature Sensors in CMOS Technology*, ACSP · Analog Circuits and Signal Processing,
https://doi.org/10.1007/978-3-030-95284-6_2

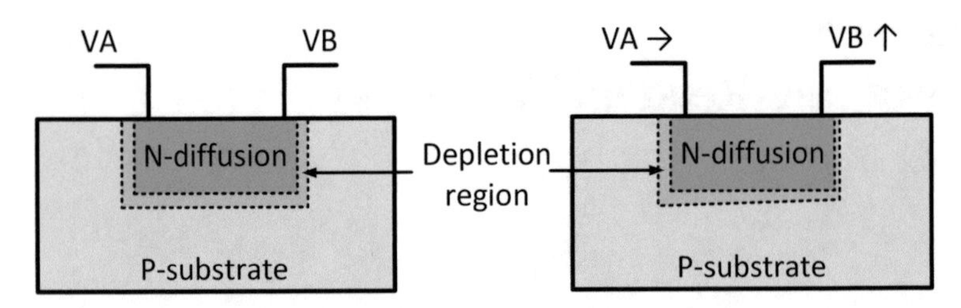

Fig. 2.1 Simplified cross section of an N+ diffusion resistor and illustration of its voltage dependency

In terms of sensitivity, metal resistors (with TCs ranging from 0.3% to 0.4%/°C) are the best choice. However, as they are optimized for high-conductivity interconnections, their sheet resistance (<100 mΩ/□) is extremely low. This results in either high power consumption or a large chip area.

The sheet resistance of diffusion resistors is typically around 100 Ω/□, which enables a compact and low-power sensor design. However, it is strongly voltage dependent. As shown in Fig. 2.1, the cross-sectional area of an N+ diffusion resistor will decrease as its potential increases and the depletion region boundary moves deeper into the N-diffusion region. Consequently, the resistor's sheet resistance will become larger. Another disadvantage of diffusion resistors is their relatively small TC: about 0.1%/°C to 0.2%/°C.

N-well resistors are very lightly doped N-diffusion resistors. Although their TCs βare large (~0.3%/°C) and comparable to that of metal resistors, their lower doping levels result in larger voltage and stress [1] sensitivities.

Unlike diffusion resistors, polysilicon resistors have a fixed geometry, and thus a much smaller voltage dependency. They also have higher sheet resistance, in the order of 100 Ω/□. In some processes, specially designed high-resistance poly resistors with sheet resistances of ~1 kΩ/□, or even higher, are also available. Depending on the technology and process, the TCs of poly resistors can be either positive or negative, ranging from −0.2%/°C to 0.1%/°C.

Compared to diffusion resistors, however, one disadvantage of poly resistors is their relatively high levels of flicker noise (1/*f* noise). As shown in Fig. 2.2, poly resistors are composed of small crystal structures which are separated by grain boundaries. These boundaries create extra energy states. When charge carriers move across such boundaries, some are trapped and later released by these states [2]. Such random trapping and releasing leads to noise that has a 1/*f* type spectrum [3]. Because the amplitude of 1/*f* noise is higher at low frequencies, it cannot be effectively suppressed by filtering or averaging. Moreover, its amplitude is proportional to the resistor's current [3], which means that increasing its power dissipation will not improve its SNR. Thus, 1/*f* noise represents a fundamental limit on the resolution of resistor-based sensors.

Another disadvantage of poly resistors is their significant long-term drift. This is mainly due to the presence of weakly bonded hydrogen atoms in their grain-boundaries [4]. Due to excessive local heating or current, these bonds may break

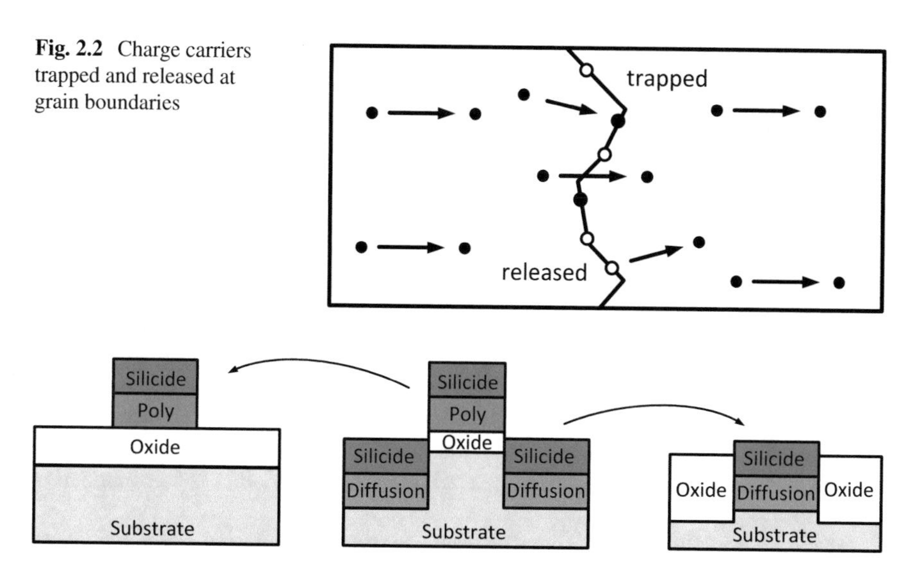

Fig. 2.2 Charge carriers trapped and released at grain boundaries

Fig. 2.3 Simplified cross section of silicided poly (left) and silicided diffusion (right) resistors compared to that of a MOSFET (middle)

and then permit traps, that is, energy states, to reform. As a result, the probability of carriers flowing through grain boundaries decreases, and so resistance increases. After being exposed to 150 °C for 6 months, resistance drifts of between 0.2% and 0.8% were observed [5]. In comparison, the drift of diffusion resistors in the same technology is 4–8 times smaller [6], and that of metal resistors is negligible.

Two other types of resistors are available in most standard CMOS technologies: silicided poly and silicided diffusion resistors. Although they are not always modelled, there are always equivalent structures used to reduce the gate/source/drain resistance of MOSFETs, as shown in Fig. 2.3. Silicide layers are formed on top of polysilicon (gate) or diffusion (source or drain) layers by first depositing a thin transition metal layer and then applying heat. The metal then reacts with the silicon, forming a low-resistance transition metal silicide. As a result, the characteristics of such resistors fall between those of metal and silicon resistors. Compared to silicon resistors, they have a relatively large and positive TC (~0.3%/°C), a more linear temperature dependence, and lower $1/f$ noise. However, their sheet resistance is much lower (~10 $\Omega/\square$). They also have low voltage and stress sensitivities, and are quite stable, showing no electrical degradation (e.g., hysteresis) even after being heated up to 500 °C [7]. Their long-term drift should be also much lower than that of silicon resistors. However, to the author's best knowledge, no serious reliability tests have been conducted for this type of resistor.

Since their characteristics are determined by the silicided layer, silicided poly and silicided diffusion resistors exhibit similar behavior in most respects. One subtle difference is their parasitic capacitances: silicided diffusion resistors have a large junction capacitor between the diffusion layer and its well/substrate, while silicided

Table 2.1 Resistor characteristics in standard CMOS processes

Resistor type	Metal	Diffusion	N-well	Poly	Silicided
1st-order TC	Large	Medium	Large	Medium or small	Large
1st-order TC sign	+	+	+	−/+	+
2nd-order TC	Medium	Medium	Large	Medium	Small
Sheet resistance	Very small	Large	Large	Large	Small
Supply dependency	Small	Medium	Large	Small	Small
1/*f* noise	Negligible	Negligible	Negligible	Large	Small or negligible
Stress sensitivity	Small	Large	Very large	Medium	Small
Drift	Very small	Small	Small	Large	Very small or small

poly resistors do not have this issue. As a consequence, silicided diffusion resistors are less suitable for high-frequency circuits.

The performance of the commonly available CMOS resistors is summarized in Table 2.1.

At the start of this research, most resistor-based temperature sensors employed diffusion or poly sensing resistors [8–11]. However, as shown above, silicided resistors are better candidates. Their only disadvantage is a relatively small sheet resistance, which can be tolerated in frequency compensation applications, where the sensing resistor is typically not very large (~100 kΩ) in order to achieve low thermal noise and high sensing resolution.

2.2.2 *Impedance Reference*

2.2.2.1 Reference Choices

In principle, a sensing resistor is a temperature-dependent impedance. Therefore, in order to digitize this impedance, a reference impedance is required.

The simplest reference impedance is a reference resistor. As illustrated in Fig. 1.11, temperature can then be sensed by reading the voltage output of a resistive divider. Alternatively, capacitors and inductors can be used to generate an impedance reference in the presence of a fixed frequency reference, as summarized in Table 2.2.

The drawbacks of an on-chip inductor reference are its large area and a low quality factor (~10) at GHz frequencies. Moreover, the quality factor is frequency dependent and drops significantly at sub-GHz frequencies. However, designing precision readout circuits operating at GHz frequencies is extremely challenging.

Capacitors are more suitable for low-frequency readout circuits. As shown in Fig. 2.4, there are three main types of on-chip capacitors: fringe capacitors (MOM, or metal-oxide-metal capacitors), plate capacitors (MIM, or metal-insulator-metal capacitors), and MOS capacitors utilizing the capacitance of the MOSFET's gate.

Benefiting from their thin gate oxide, MOS capacitors have a high capacitance density. However, their capacitance is voltage dependent, as the space charge region

Table 2.2 Characteristics of different references

Type	Resistor	Capacitor	Inductor
Symbol			
V-I characteristic	$V = RI$	$I = C \cdot \frac{dV}{dt}$	$V = L \cdot \frac{dI}{dt}$
Impedance	R	$\frac{1}{j\omega C}$	$j\omega L$
Structure	Poly, diffusion, metal, etc.	Fringe, plate, MOS gate	Planar
Quality factor	–	High	Low

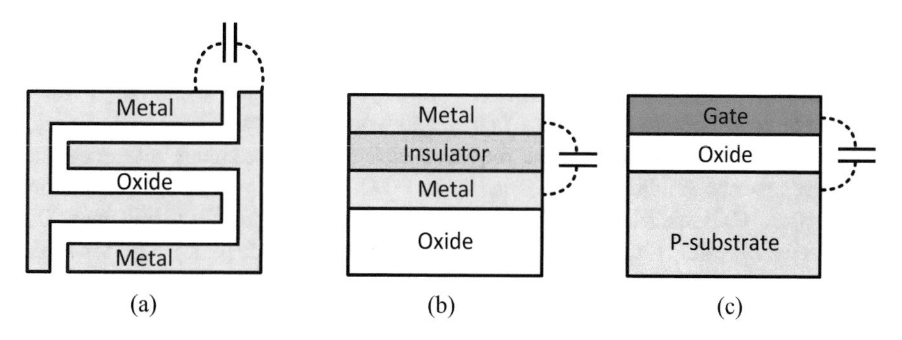

Fig. 2.4 (**a**) Top view of a MOM capacitor. (**b**) Cross section of a MIM capacitor. (**c**) Cross section of a MOS capacitor

of the substrate is modulated by the gate voltage. Additionally, the MOS capacitor is sensitive to temperature and process spread: when used as the impedance reference, the resistor-based sensor's accuracy will be degraded.

MIM and MOM capacitors are much more stable due to their well-defined electrode spacing and the low temperature dependency of their oxide dielectrics. Typically, their TC is smaller than 100 ppm/°C [12]. In some modern CMOS processes, the density of MOM capacitors is higher than that of MIM capacitors. However, in mature processes, for example, 0.18 μm CMOS, the MIM capacitor still has a higher density, and is thus preferred in area-constrained designs.

The drift of capacitors is mainly caused by electrical stress [13, 14], which creates trapped charges in the oxide insulator layer. This, in turn, can generate new dipoles and so modulate the dielectric permittivity. As this is determined by the strength of the electrical field on the dielectric, operating the circuit at standard supply voltages should almost eliminate this effect. In [15], the 1fF/μm^2 MIM capacitor drifts by less than 0.01% with an applied voltage of 5 V. This implies that, for most MIM capacitors with densities ≤2fF/μm^2, no observable drift will occur if the applied voltages are kept below 2.5 V.

2.2.2.2 Comparison

As discussed above, both resistors and capacitors are suitable references for resistor-based sensors. Both structures have their own advantages and disadvantages.

The benefits of using a resistive reference include ease of use and high efficiency. Furthermore, it does not need to operate at a well-defined frequency. Advantageously, the *TC* of the reference resistor may even be chosen to be opposite to that of the sensing resistor, resulting in a resistor divider with increased sensitivity. Assuming that the TC_s of two resistors are TC_p (positive) and TC_n (negative), the theoretical FoM of such a resistor divider sensor (Fig. 1.11) can be expressed as:

$$FoM_{\mathrm{WhB}} = \frac{8kT}{\left(TC_{\mathrm{p}} - TC_{\mathrm{n}}\right)^2}, \tag{2.1}$$

which is smaller (better) than that given in Eq. (1.13).

However, as shown in Table 2.1, only certain poly resistors have a negative TC, in contrast to the positive TC of silicided sensing resistors. As a result, the price for better energy efficiency is excess 1/*f* noise and worse stability.

RC sensors, on the other hand, require a well-defined frequency reference and are less energy efficient. However, as the TC of MIM or MOM capacitors (<100 ppm/°C) is much lower compared to that of resistors (typically >1000 ppm/°C), these sensors suffer from less spread and can achieve higher accuracy with the same number of trimming points. Also, compared to dual-R sensors with polysilicon reference resistors, RC sensors implemented with silicided resistors are more stable.

The features of dual-R and RC temperature sensors are summarized in Table 2.3.

2.2.3 *Sensor Structures and Readout Method*

2.2.3.1 Dual-R Sensors

There are several ways to digitize the ratio between two resistors with different TCs (R_p and R_n). Compared to the simple resistor divider (Fig. 2.5a), the use of current sources allows for a small differential output that can be easily processed (Fig. 2.5b) [8].

Table 2.3 Resistor-based sensor overview

Type	Dual-R	RC
Reference	Resistor	Capacitor + frequency
Accuracy	Medium	High
Efficiency	High	Medium
Stability	Medium	High

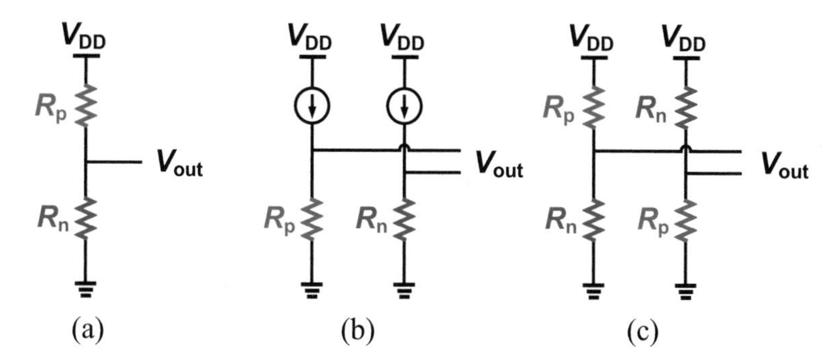

Fig. 2.5 Several dual-R sensor structures. (**a**) Resistor divider. (**b**) Current-driven resistors. (**c**) Wheatstone bridge

However, the excess noise produced by the current source will limit noise efficiency. The most commonly used structure is the fully differential Wheatstone bridge (WhB) (Fig. 2.5c). Apart from improved power supply sensitivity [15], it does not require matched current sources.

Other than the voltage readout methods shown in Fig. 2.5, dual-R sensors can be configured to output a temperature-dependent current. The advantages and disadvantages of the two approaches will be discussed in more detail in Chap. 4.

2.2.3.2 RC Sensor Structures

Based on how they are driven by a frequency reference, RC sensors can be divided into two main categories: discrete-time and continuous-time. Just like dual-R sensors, their power supply sensitivity can be improved by using a fully differential topology. In the following sections, however, simple single-ended schematics will be used to illustrate their working principles.

One type of discrete-time RC sensor employs switched-capacitor resistor references, as shown in Fig. 2.6a [16]. Given a fixed input frequency f_s, the effective reference resistance R_0 can then be expressed as $R_0 = 1/(C_0 f_s)$. By using a large capacitor C_F to filter out switching transients, the resulting voltage output is similar to that of a real resistor divider. Figure 2.6b shows an alternative topology based on the settling characteristics of a low-pass RC filter [17]. The capacitor C_0 is first reset to V_{DD} (φ_{RST}), while during the next phase (φ_{DCHG}), C_0 is discharged through the sensing resistor R_S. After a fixed period, the discharging stops, and the residual voltage on C_0 is a representation of the RC time constant and temperature.

As already mentioned, one advantage of RC-based sensors over dual-R sensors is high accuracy. However, the switching operation of discrete-time sensors is accompanied by charge injection, which causes errors in the voltage stored on the capacitors. As a result, the sensor becomes less accurate.

Charge injection can be minimized if its current path to the reference capacitor is blocked by a resistor. As a result, temperature sensors built around continuous-time

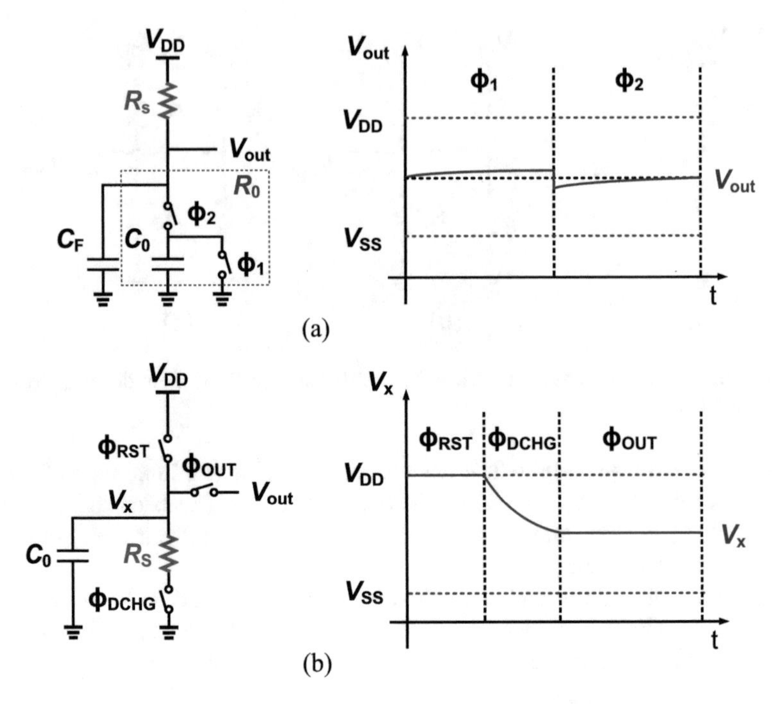

Fig. 2.6 Discrete-time RC sensors: (**a**) Using a switched-capacitor resistor. (**b**) Using incomplete settling of an RC filter

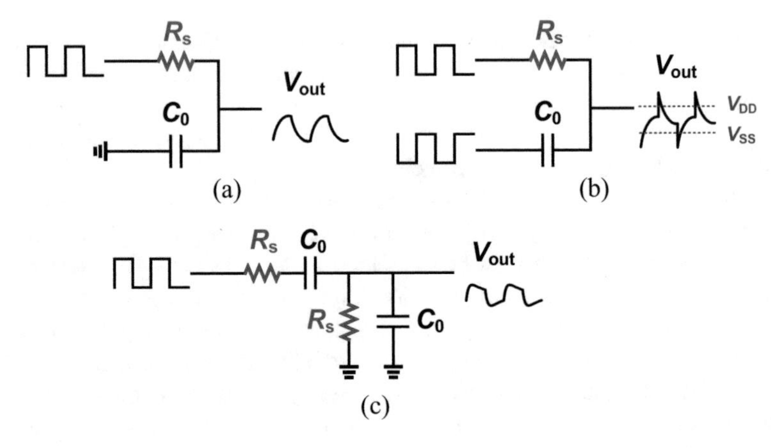

Fig. 2.7 Continuous-time RC sensors: (**a**) Low-pass filter. (**b**) Poly-phase filter. (**c**) Wien bridge (WB) filter

RC filters are potentially more accurate. Some of the filter variations, including the low-pass filter, poly-phase filter [18], and Wien bridge (WB) [10, 11], are shown in Fig. 2.7. Given a fixed frequency input, which is typically a square wave generated

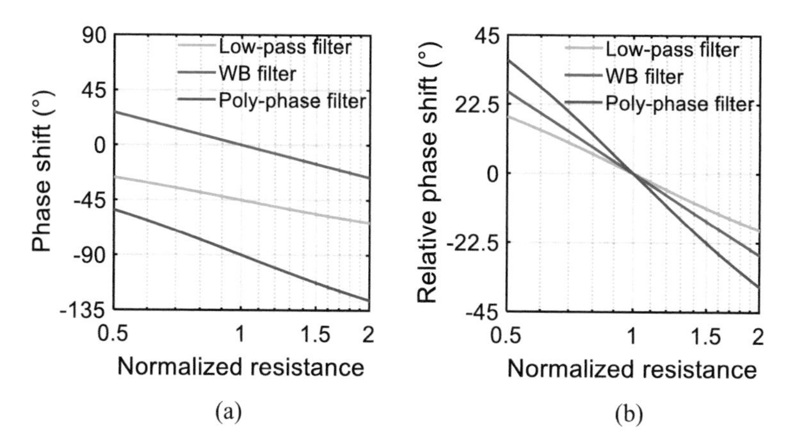

Fig. 2.8 (**a**) Phase shift and (**b**) relative phase shift of different RC filters as a function of normalized resistance

from inverters, the phase shift of the output waveform is determined by the temperature-dependent RC time constant.

Assuming all the filters are operated around their center frequencies ($\omega R_s C_0 \approx 1$), their phase shifts as a function of normalized resistance are presented in Fig. 2.8a. Of the three, the poly-phase filter has the highest phase sensitivity [18] (Fig. 2.8b), and thus the best theoretical FoM. At the center frequency, its phase sensitivity with respect to R_s is $1/R_s$, which is 2×/1.5× larger than that of the low-pass filter/Wien bridge filter, respectively. However, when driven by a rail-to-rail square wave, the output voltage of the poly-phase filter will exceed the supply rails, thus imposing a stringent requirement on the input stage of the readout circuit. Moreover, high-frequency supply noise can be directly coupled to its output signal. On the other hand, the Wien bridge filter has the second-largest sensitivity and an inherent filtering of high-frequency noise. Thus, it is the preferred building block of high-accuracy temperature sensors.

2.2.3.3 RC Filter Readout

An RC filter can be seen as a circuit that outputs different phases at different input frequencies. As such, there are two ways to sense its RC time constant: by fixing its input frequency and extracting the output phase, or by enforcing a certain output phase and measuring the required excitation frequency. Extracting its RC time constant by sensing its output amplitude is also possible. However, if not processed carefully, the result will be highly sensitive to the supply voltage.

The former method is depicted in Fig. 2.9a. Typically, it requires a reference clock that generates two signals of the same frequency: a driving signal φ_{dirve}, and a reference signal φ_{ref} which serves as the input of a phase-ADC. As will be discussed in detail in Chap. 3, this usually consists of two blocks: a phase detector, which

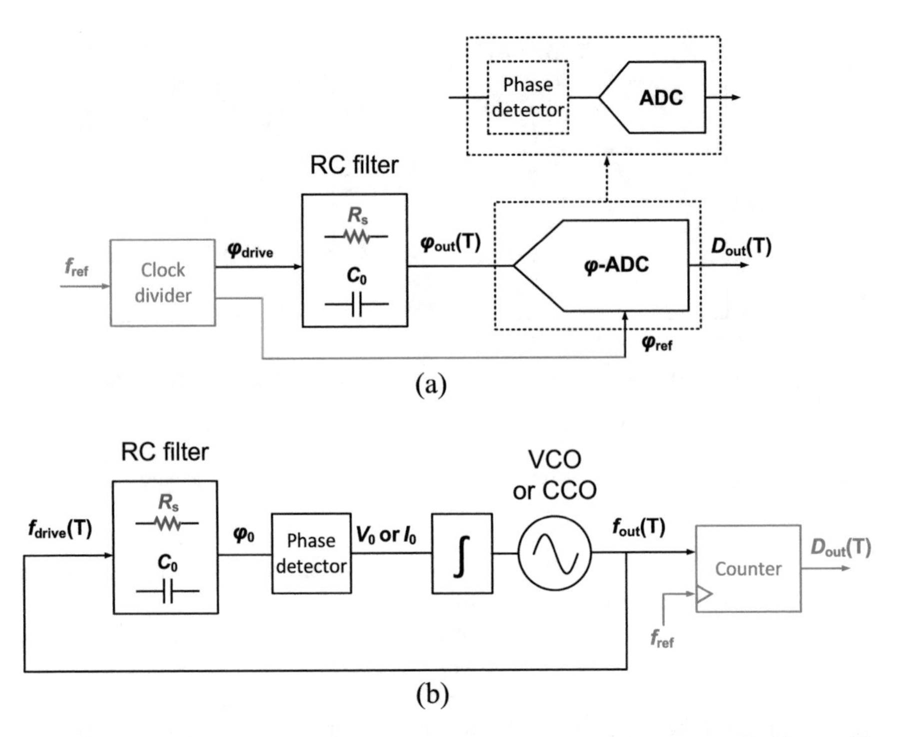

Fig. 2.9 RC filter readout methods using (**a**) phase-ADC and (**b**) frequency-locked loop, with elements in green indicating the generation and processing of clock signals from an external reference

converts the filter's phase shift into a voltage or current, followed by a conventional amplitude-domain ADC.

The latter method can be implemented by a frequency-locked-loop (FLL) [11, 18], as shown in Fig. 2.9b. A phase detector first converts the phase output of the RC filter into a voltage or current signal, which, after integration, controls the driving frequency of the RC filter. The feedback loop then forces the output of the phase detector to zero, thus fixing the RC filter's output phase, and making the VCO's output frequency inversely proportional to the filter's RC time constant. Since an FLL outputs an RC-dependent frequency, a frequency reference is still required for digitization, usually with the help of a digital counter.

Both structures have their advantages and disadvantages. The phase-ADC approach is simple and direct. However, the shape of the RC filter's output waveform will vary over temperature, imposing stringent nonlinearity requirements on the phase-ADC in order to obtain accurate sensor readout. The FLL-based readout, on the other hand, does not have this issue. However, its sensing resolution is typically limited by the counter's clock frequency. For example, to obtain a 0.5 mK quantization-noise-limited resolution from an RC filter with a 0.3%/°C TC, the counter should count up to over 1/(0.3%·0.0005) times. Assuming a 5 ms conversion time, the reference clock frequency should be greater than 133 MHz. As a result, the counter will be very power hungry, especially in mature processes (e.g., 0.18 μm CMOS), which limits the achievable energy efficiency of such sensors. Thus, the phase-ADC is better suited for the high-resolution task of compensating the temperature variations of on-chip frequency references.

2.3 ADC Choice

As discussed above, the FLL-based readout of RC-based sensors is not well suited to the target application, so a phase ADC will be used. This in turn requires an amplitude-domain ADC. There are mainly three requirements for the ADC. First, to ensure sufficient sensing resolution, the ADC should have a high SNR. For example, achieving 0.5 mK resolution in a 200 °C temperature range requires an SNR of >112 dB. Second, to achieve a high energy efficiency of the overall circuit, the ADC should have a balanced power/noise performance compared to the sensor front-end. This means that the ADC should be as efficient as possible, for example, have a good Schreier FoM [19]. Last but not least, the ADC should have high linearity and accuracy, so that the sensor's trimming effort can be minimized.

2.3.1 *Nyquist Versus Oversampled ADCs*

Many sensor readout circuits use Successive Approximation ADCs (SAR ADCs) because of their low power. However, due to the binary nature of their DAC elements, errors like nonmonotonicity and missing-codes limit their resolution. As a result, this is typically limited to ~14 bits, or a ~85 dB SNR, as shown in Fig. 2.10 [20]. With state-of-the-art background calibration, their SNR can be improved to about 100 dB [21]. However, this is still below the stringent requirement (>110 dB) imposed by the targeted temperature sensing application. Another major type of Nyquist ADCs is the pipeline ADC. Although these can be extremely fast, their resolution is typically worse than that of a SAR ADC: other than DAC mismatch, additional errors are introduced during the generation of the inter-stage residue signals.

Oversampled ADCs (ΔΣ-ADCs) can achieve low quantization noise levels by shaping quantization noise away from a bandwidth of interest, and then suppressing it through filtering. This enables the use of inherently linear 1-bit quantizers and DACs. If multi-bit DACs are used, their mismatch errors can also be shaped by using algorithms like data-weighted averaging (DWA) [22]. As a result, oversampled ADCs can achieve higher SNR than Nyquist ADCs, and meet the resolution requirements of the temperature compensation application. Furthermore, they can also achieve excellent energy efficiency [20].

Due to oversampling, ΔΣ-ADCs have a lower conversion rate than Nyquist ADCs. However, this is not an issue in the targeted application, which only requires bandwidths of about 100 Hz.

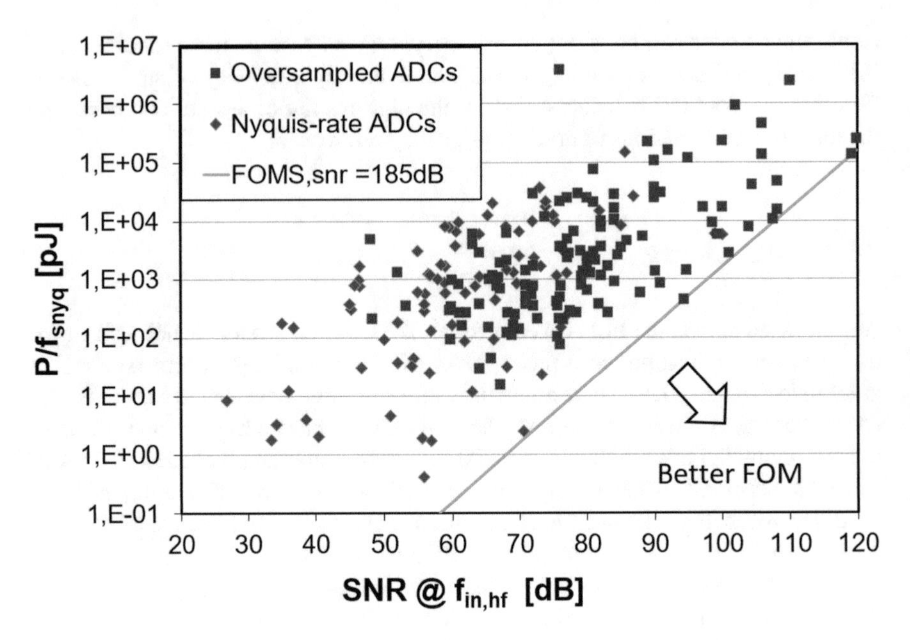

Fig. 2.10 Power per bandwidth versus SNDR of Nyquist-rate and oversampled ADCs, with a constant Schreier FoM line [20]

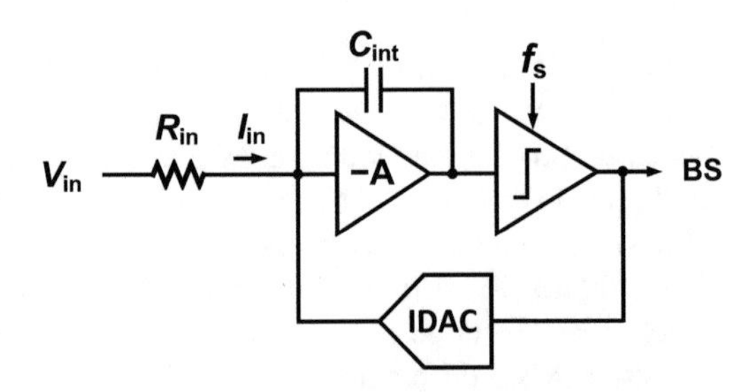

Fig. 2.11 Basic structure of a CT ΔΣ-ADC

2.3.2 *Continuous-Time ΔΣ-ADC*

Reading out a resistor always involves passing a current through it. As a result, a continuous-time (CT) ΔΣ-ADC (Fig. 2.11) is the most natural choice. As will be shown in the following chapters, the temperature sensing resistor(s) can be reconfigured as the input resistor (R_{in}) in a CT ΔΣ-ADC, so the current flowing through the sensing resistor can be directly digitized.

Table 2.4 Major design choices of resistor-based temperature sensors

Sensing resistor	Silicided resistor	
Reference (sensor type)	Poly resistor (dual-R)	Capacitor + frequency (RC)
Advantage/focus	High efficiency	High accuracy
Preferred sensor structure	Wheatstone bridge	Wien bridge
Readout circuit	CT ΔΣ-ADC	

2.4 Concluding Remarks

In this chapter, major design choices for both the sensor front-end and the readout circuit have been made based on the requirements of the targeted frequency compensation application. These are summarized in Table 2.4. Of the available CMOS resistors, silicided resistors are the best choice due to their high sensitivity and high stability. Depending on application requirements, the reference impedance can be achieved using either a negative-TC poly resistor for high energy efficiency, or a stable low-TC capacitor driven at a reference frequency for high accuracy. For the dual-R/RC type of sensor, a Wheatstone bridge/Wien bridge is preferred. As for the readout circuit, the resolution requirement can only be achieved by using oversampled ADCs. Due to the resistive sensing requirement, CT ΔΣ-ADC becomes a natural choice.

Having made these design choices, the following two chapters will focus on the design and characterization of both Wien bridge (Chap. 3) and Wheatstone bridge (Chap. 4) sensor prototypes.

References

1. O.N. Tufte, P.W. Chapman, D. Long, Silicon diffused-element piezoresistive diaphragms. J. Appl. Phys. **33**(11), 3322–3327 (1962)
2. B. Razavi, *Design of Analog CMOS Integrated Circuits* (Tata McGraw-Hill Education, 2002)
3. H.C. De Graaff, M.T.M. Huybers, 1/*f* noise in polycrystalline silicon resistors. J. Appl. Phys. **54**(5), 2504–2507 (1983)
4. M. Rydberg, U. Smith, Long-term stability and electrical properties of compensation doped poly-Si IC-resistors. IEEE Trans. Electron. Devices **47**(2), 417–426 (2000)
5. A. Andrei, C. Malhaire, S. Brida, D. Barbier, Reliability study of AlTi/TiW, polysilicon and ohmic contacts for piezoresistive pressure sensors applications, in *IEEE SENSORS*, vol. 3, (2004), pp. 1125–1128
6. S. Jose, J. Bisschop, V. Girault, L.V. Marwijk, J. Zhang, S. Nath, Reliability of integrated resistors and the influence of WLCSP bake, in *IEEE Proc. IIRW*, (2016), pp. 69–72
7. E. Vereshchagina, R.A.M. Wolters, J.G.E. Gardeniers, The development of titanium silicide–boron-doped polysilicon resistive temperature sensors. J. Micromech. Microeng. **21**(10), 105022 (2011)
8. Z. Tang, Y. Fang, X.-P. Yu, Z. Shi, L. Lin, N.N. Tan, A dynamic-biased resistor-based CMOS temperature sensor with a duty-cycle-modulated output. IEEE Trans. Circuits Syst. II **67**(9), 1504–1508 (2020)

9. C. Weng, C. Wu, T. Lin, A CMOS thermistor-embedded continuous-time Delta-sigma temperature sensor with a resolution FoM of 0.65 pJ °C^2. IEEE J. Solid-State Circuits **50**(11), 2491–2500 (2015)
10. M. Shahmohammadi, K. Souri, K.A.A. Makinwa, A resistor-based temperature sensor for MEMS frequency references, in *Proc. ESSCIRC*, (2013), pp. 225–228
11. P. Park, D. Ruffieux, K.A.A. Makinwa, A thermistor-based temperature sensor for a real-time clock with ±2 ppm frequency stability. IEEE J. Solid-State Circuits **50**(7), 1571–1580 (2015)
12. Q.S.I. Lim, A.V. Kordesch, R.A. Keating, Performance comparison of MIM capacitors and metal finger capacitors for analog and RF applications, in *RF and Microwave Conf.*, (2003), pp. 85–89
13. C. Besset et al., MIM capacitance variation under electrical stress. Microelectron. Reliab. **43**(8), 1237–1240 (2003)
14. C. Hung et al., An innovative understanding of metal–insulator–metal (MIM)-capacitor degradation under constant-current stress. IEEE Trans. Device Mater. Reliab. **7**(3), 462–467 (Sept. 2007)
15. J. Van Rethy, H. Danneels, V. De Smedt, W. Dehaene, G.E. Gielen, Supply-noise-resilient design of a BBPLL-based force-balanced Wheatstone bridge interface in 130-nm CMOS. IEEE J. Solid-State Circuits **48**(11), 2618–2627 (2013)
16. M.H. Perrott et al., A temperature-to-digital converter for a MEMS-based programmable oscillator with <±0.5-ppm frequency stability and <1-ps integrated jitter. IEEE J. Solid-State Circuits **48**(1), 276–291 (2013)
17. A. Khashaba et al., A 0.0088mm^2 resistor-based temperature sensor achieving 92fJ·K^2 FoM in 65nm CMOS. IEEE ISSCC Dig. Tech. Papers, 60–61 (2020)
18. W. Choi et al., A compact resistor-based CMOS temperature sensor with an inaccuracy of 0.12°C (3σ) and a resolution FoM of 0.43pJ·K^2 in 65-nm CMOS. IEEE J. Solid-State Circuits **53**(12), 3356–3367 (2018)
19. R. Schreier, G.C. Temes, *Understanding Delta-Sigma Data Converters* (Wiley, New York, 2005)
20. B. Murmann, ADC Performance Survey 1997–2020. [Online]. Available: http://web.stanford.edu/~murmann/adcsurvey.html
21. H. Li, M. Maddox, M.C.W. Coin, W. Buckley, D. Hummerston, N. Naeem, A signal-independent background-calibrating 20b 1MS/S SAR ADC with 0.3ppm INL, in *IEEE ISSCC Dig. Tech. Papers*, (2018), pp. 242–244
22. R.T. Baird, T.S. Fiez, Improved $\Delta\Sigma$ DAC linearity using data weighted averaging. Proc. ISCAS **1**, 13–16 (1995)

Chapter 3
Wien Bridge–Based Temperature Sensors

3.1 Introduction

In Chap. 2, RC and dual-R based sensors were presented and compared. RC-based sensors rely on a stable reference capacitor, which makes them suitable for building accurate temperature sensors. This chapter discusses a particular and suitable approach: the readout of a Wien bridge (WB) sensor with a phase-domain ADC. It begins with a discussion of general design choices. Then, three different WB sensor prototypes are presented based on their publication sequence.

3.2 General Design Choices

3.2.1 WB Sensor

The circuit diagram of a differential WB is shown in Fig. 3.1a. It is a second-order band-pass filter whose voltage amplitude and phase transfer functions can be expressed, respectively, by eqs. 3.1 and 3.2.

$$\mathrm{H}(j\omega) = \frac{R_s C_0 j\omega}{1 - R_s^2 C_0^2 \omega^2 + 3R_s C_0 j\omega} \tag{3.1}$$

$$\varphi_{\mathrm{WB}}(\omega) = -\tan^{-1}\left(\frac{R_s^2 C_0^2 \omega^2 - 1}{3R_s C_0 j\omega}\right) \tag{3.2}$$

S. Pan, K. A. A. Makinwa, *Resistor-based Temperature Sensors in CMOS Technology*, ACSP · Analog Circuits and Signal Processing,
https://doi.org/10.1007/978-3-030-95284-6_3

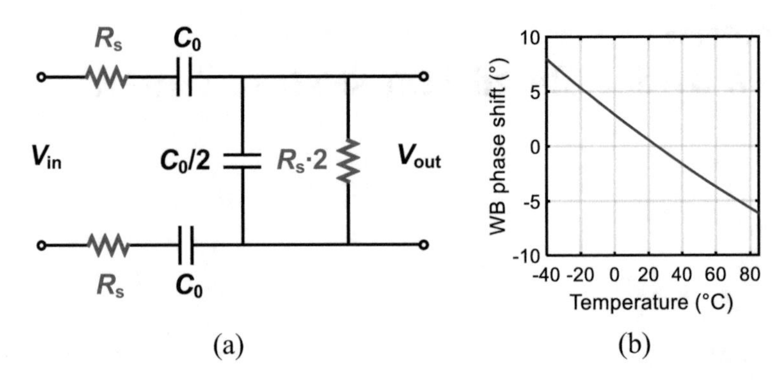

Fig. 3.1 (**a**) Differential Wien-Bridge and (**b**) its phase output as a function of temperature.

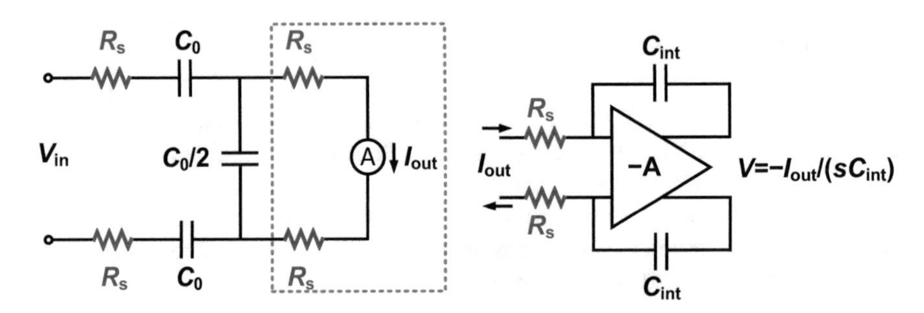

Fig. 3.2 (**a**) Current-based readout of a WB. (**b**) Sensing resistors reused as input resistors in a CT ΔΣ-ADC

Given a temperature-dependent RC time-constant but a fixed input frequency (at around the WB's center frequency), the amplitude of the WB output will remain roughly constant, while the output phase will vary over temperature.

As discussed in Chap. 2, to reduce chip area, the WB capacitors (C_0) should be realized as high-density metal-insulator-metal (MIM) capacitors, and to reduce supply sensitivity, silicided poly resistors (less parasitic capacitance) are preferred over silicided diffusion resistors. The temperature dependence of the WB phase will then be mainly determined by the silicided poly resistor, since the TC of MIM capacitors is quite low (<100 ppm/°C). In the chosen 0.18 μm technology, the resistor's TC is about 0.285%/°C, and so the WB phase will vary by about 15° over the industrial temperature range (−40 °C to 85 °C), as shown in Fig. 3.1b.

The same temperature-dependent phase response can be obtained by measuring the current through the output resistors [1], as shown in Fig. 3.2a. This facilitates the readout of a WB by a CT ΔΣ-ADC, as two of the WB resistors can be reused as the ADC's input resistors (Fig. 3.2b). The WB phase can then be detected by inserting a phase detector in the integrator's feedback loop, as will be discussed in Sect. 3.2.2.

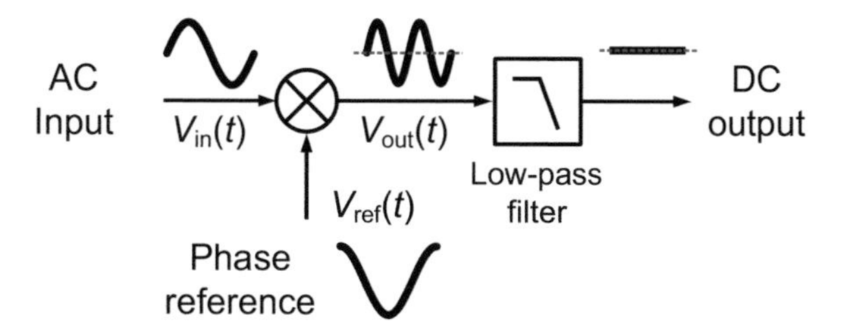

Fig. 3.3 A phase detector built with an analog multiplier and a low-pass filter

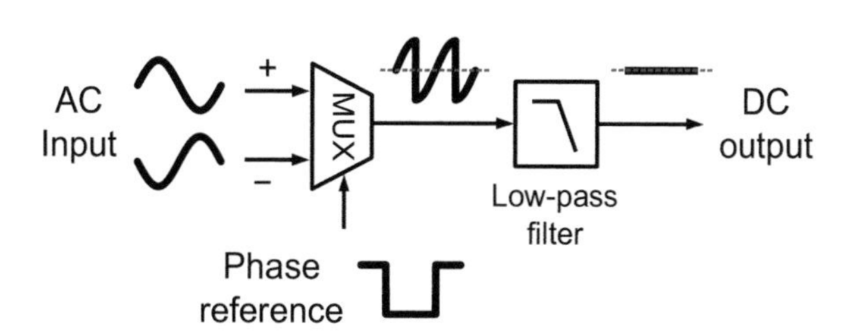

Fig. 3.4 Phase detector built with an analog multiplexer and a low-pass filter

3.2.2 *Phase-Domain ADC*

3.2.2.1 Phase Detector

Phase detection is essential to the operation of a phase-domain ADC. For analog input signals, it can be achieved with the help of an analog multiplier and a low-pass filter, as shown in Fig. 3.3. Here, both the input signal and the reference signal are sine-wave signals with the same frequency f_0, that is,

$$V_{in}(t) = A_{in} \cdot \sin\left(2\pi f_0 t + \varphi_{in}\right)$$

$$V_{ref}(t) = A_{ref} \cdot \sin\left(2\pi f_0 t + \varphi_{ref}\right) \tag{3.3}$$

The multiplier's output can then be expressed as:

$$V_{out}(t) = \frac{A_{in} A_{ref}}{2} \cdot \left(\cos\left(\varphi_{in} - \varphi_{ref}\right) - \cos\left(4\pi f_0 t + \varphi_{in} + \varphi_{ref}\right)\right). \tag{3.4}$$

It contains two terms, the first is a DC signal that contains the phase difference information, while the second contains high-frequency content that can be removed by a low-pass filter. Due to the cosine function, the DC signal achieves the maximum sensitivity and linearity when $\varphi_{in} - \varphi_{ref} \approx 90°$.

Alternatively, the reference signal may be a square-wave, allowing the analog multiplier to be replaced by an analog multiplexer, as shown in Fig. 3.4. Depending on the value of the phase reference, the multiplexer outputs either the original or the inverted version of the input signal. This precisely realizes a reference square-wave signal with an amplitude $A_{ref} = 1$. As for a sine-wave reference, a low-pass filter can be used to eliminate the higher-order multiplication products, leaving a DC signal that represents the desired phase information.

3.2.2.2 Phase DAC and Phase-Domain ΔΣ-ADC

Although the multiplier approach (Fig. 3.3) is quite straightforward, it requires an accurate and low-noise reference phase signal as well as a precise analog multiplier, which is often power-hungry. On the other hand, the multiplexer approach shown in Fig. 3.4 can be realized by switches, and it is thus much more energy efficient. In the case of a current-mode WB sensor, the output current direction can be toggled by a chopper, effectively multiplying the WB output current by a square wave of unity amplitude, as shown in Fig. 3.5. By using a multiplexer to select various phase references, the phase detector circuit can be turned into a phase DAC. Apart from the reference signal, the driving signal of the WB is also made a square wave to achieve low noise and high energy efficiency. The waveforms over temperature are plotted in Fig. 3.6.

As shown in Fig. 3.7, a phase-domain Delta-Sigma ADC (PDΔΣ-ADC) can be built around a WB sensor by incorporating a phase DAC in the feedback path of a continuous-time ΔΣ-ADC [2]. The ADC first down-converts $\varphi_{WB}(T)$ by multiplying it by a phase reference at the same frequency ($f_{demod} = f_{drive}$). Depending on the chosen references φ_0 or φ_1 of the phase DAC, the multiplier's DC output is either positive or negative [1]. The multiplier's output is filtered by the integrator and then quantized. In a negative feedback loop, the quantizer toggles the reference phases such that the loop filter's average DC input is zero. The average of the output bitstream is therefore a digital representation of $\varphi_{WB}(T)$.

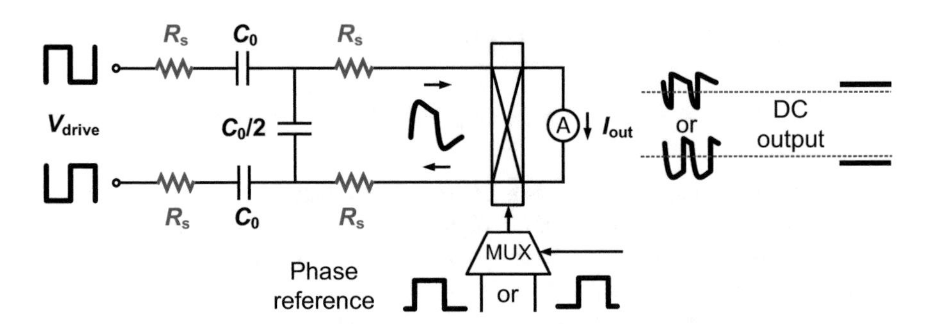

Fig. 3.5 WB sensor and analog multiplexer/phase DAC achieved by a current chopper

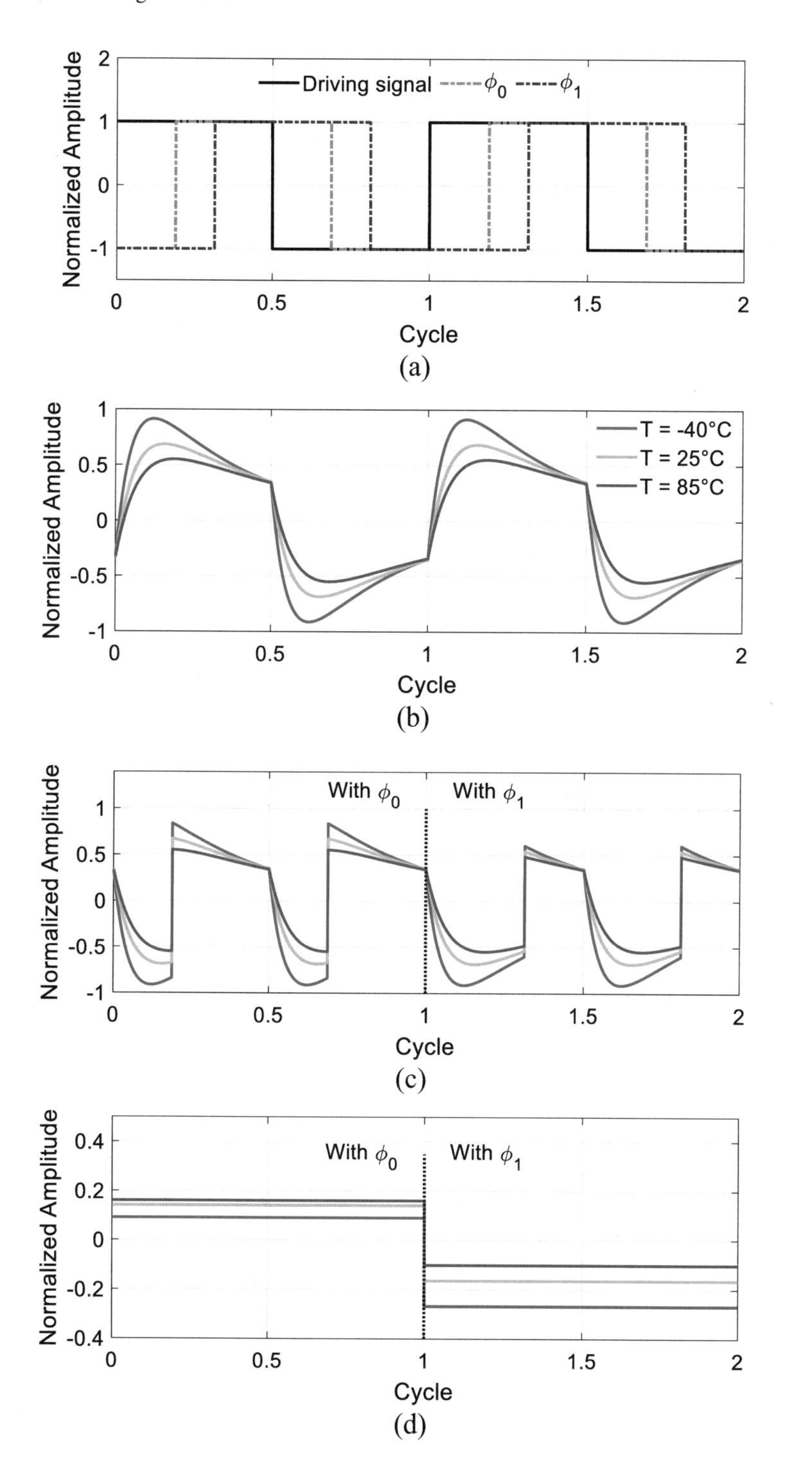

Fig. 3.6 (**a**) WB driving and phase DAC signals. (**b**) WB sensor output under different temperatures. (**c**) WB sensor output after analog multiplexer. (**d**) Averaged output over one cycle

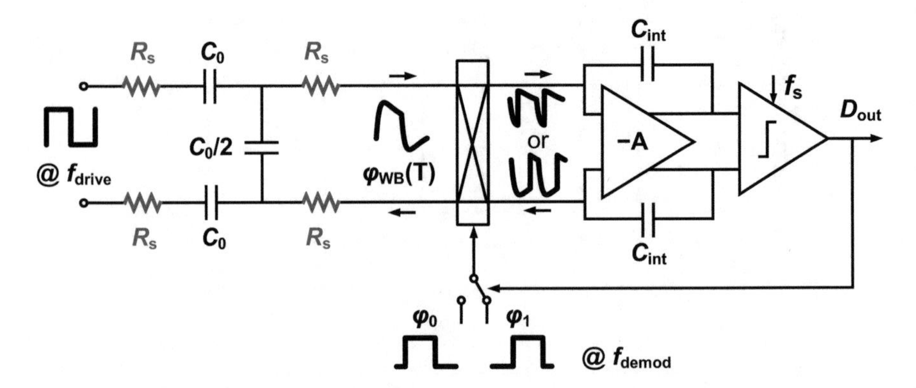

Fig. 3.7 WB sensor digitized by a phase-domain ΔΣ-ADC

3.2.3 System Analysis

3.2.3.1 Resolution and FoM

Besides the promising accuracy of WB sensors, resolution and resolution FoM [3] remain important specifications. Two simplifications are made to obtain the WB sensor's theoretical FoM, for example, the FoM without considering the power and noise of its readout circuits. First, both the driving and the demodulating signals are assumed to be sine waves with the same frequency: $V_{in} = A{\cdot}\sin(2\pi f_0 t)$ and $V_{demod} = \sin(2\pi f_0 t + \varphi_{demod})$, where A is the amplitude of the driving signal. Second, the WB filter is assumed to be driven at its center frequency, that is, $f_0 = 1/(2\pi R_S C_0)$, and $\varphi_{demod} = 90°$.

The theoretical resolution of the WB sensor can then be derived by comparing the levels of the noise and the temperature-dependent DC signal present at the output of the low-pass filter. The demodulating signal is assumed to be noise-free, and thus its amplitude does not affect the sensor's resolution.

Under the orthogonal sine-wave assumption, the sensitivity of the demodulator's DC output (I_{DC}) to temperature can be expressed as:

$$S_{WB} = \frac{dI_{DC}}{dT} = \frac{dI_{DC}}{d\varphi_{WB}} \cdot \frac{d\varphi_{WB}}{dT} = \frac{A}{6R_S} \cdot \frac{2TC_s}{3} = \frac{A \cdot TC_s}{9R_S}. \tag{3.5}$$

At the driving frequency f_0, the noise spectrum densities of R_1 and R_2 in Fig. 3.8 before demodulation are $i_{n,R1} = \sqrt{4kT/(9R_s)}$ and $i_{n,R2} = \sqrt{4kT \cdot 5/(9R_s)}$, respectively. After demodulation, the noise power will be reduced by 4×, as the power of the unity-amplitude demodulation sine wave is 0.5, and only half of the noise power will be demodulated to DC. Given a conversion time of t_{conv}, the amplitude of the output current noise becomes:

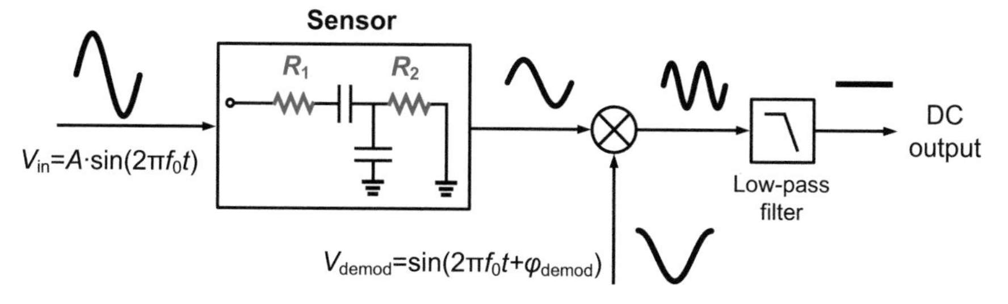

Fig. 3.8 Phase detection model for WB sensor's resolution calculation

$$I_{\mathrm{n}} = \sqrt{{i_{\mathrm{n,R1}}}^2 + {i_{\mathrm{n,R2}}}^2} \cdot \sqrt{\frac{1}{2t_{\mathrm{conv}}}} \cdot \sqrt{\frac{1}{4}} = \sqrt{\frac{kTR_{\mathrm{S}}}{3t_{\mathrm{conv}}}}. \tag{3.6}$$

By combining (3.5) and (3.6), the resolution of this WB sensor can be expressed as:

$$\Delta T = \frac{I_{\mathrm{n}}}{\mathrm{S_{WB}}} = \frac{3}{A \cdot TC_{\mathrm{S}}} \sqrt{\frac{3kTR}{t_{\mathrm{conv}}}}. \tag{3.7}$$

For a differential WB sensor, however, the resolution improves by $\sqrt{2}$ due to the halved noise power, that is,

$$\Delta T_{\mathrm{diff}} = \frac{3}{A \cdot TC_{\mathrm{S}}} \sqrt{\frac{3kTR}{2t_{\mathrm{conv}}}}. \tag{3.8}$$

When driven at its center frequency f_0, the power consumed by the differential bridge is $A^2/(3R_S)$. By combining this with (3.8), the sensor's FoM can be calculated as:

$$\mathrm{FoM_{WB}} = \frac{9kT}{2 \cdot {TC_{\mathrm{S}}}^2}. \tag{3.9}$$

Assuming $TC_S = 0.285\%/°C$, the theoretical resolution FoM is about 2.3 fJ·K^2.

In practice, it is easier and more accurate to use square-wave drive and demodulation. In this case, the higher-order harmonics will reduce the phase sensitivity and increase the power consumption of the WB. Simulations show that the theoretical FoM will then degrade to 9.7 fJ·K^2. In a real design, however, the power and noise of the readout circuit will further degrade the FoM.

3.2.3.2 Nonlinearity and Trimming

To reduce the calibration costs of a temperature sensor, the number of trimming points should be minimized. In the case of a WB sensor, this is limited by the spread of the sensing resistor. As presented in Chap. 1, a 2-point trim is required to achieve good accuracy. Thus, the readout circuit should not introduce extra error after a 2-point trim.

Unfortunately, applying a 2-point trim to the bitstream average of a phase-domain ΔΣ-ADC will result in extra spread. This is because, due to RC spread, nominally identical WB sensors will output slightly different phase shifts. Since the readout system is nonlinear, the transfer function of each sample will then be slightly different, as shown in Fig. 3.9. As a result, the residual nonlinearity will also be slightly different over samples, and so more trimming points (>2) will be required to achieve high accuracy.

To avoid increasing the number of trimming points, the nonlinearity of the sensor readout should be compensated before trimming. This nonlinearity mainly consists of two parts: (a) the nonlinear RC-to-phase characteristic of the RC filter and (b) the nonlinearity of the phase-domain ΔΣ ADC. Fortunately, both of these are fully deterministic and so can be corrected mathematically.

Under a sine-wave excitation/demodulation assumption, the so-called cosine nonlinearity of a phase-domain ΔΣ ADC can be readily computed [1, 2]. At steady state, the DC input of the ΔΣ-ADC's integrator has an average value of zero. Thus, the WB phase φ_{WB} and the ΔΣ-ADC's bitstream average μ are related by:

$$\frac{\mu A}{2}\cdot\cos\left(\varphi_1-\varphi_{\mathrm{WB}}\right)+\frac{\left(1-\mu\right)A}{2}\cdot\cos\left(\varphi_1-\varphi_{\mathrm{WB}}\right)=0. \tag{3.11}$$

From this equation, φ_{WB} is clearly a nonlinear function of μ. After some manipulation, φ_{WB} can be expressed analytically as:

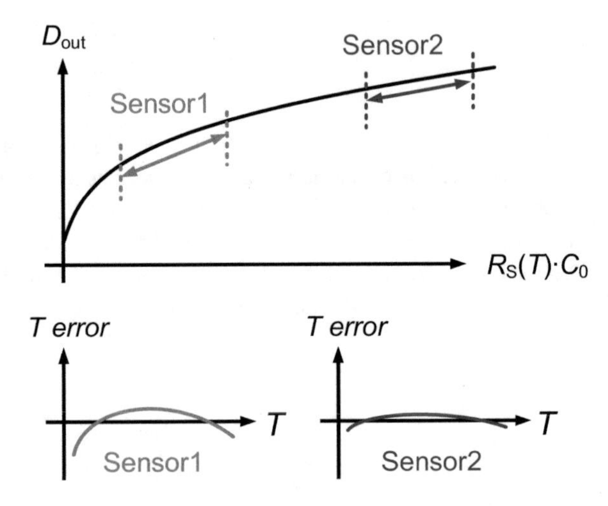

Fig. 3.9 Illustration of the effect of readout nonlinearity on sensor's accuracy after a 2-point trim

$$\varphi_{\mathrm{WB}} = \tan^{-1}\left(\frac{\mu\cos(\varphi_1 - \varphi_0) - \mu + 1}{\mu\sin(\varphi_1 - \varphi_0)}\right) - \varphi_0. \tag{3.12}$$

As shown in Fig. 3.10, the relative error due to the cosine nonlinearity decreases rapidly as the phase DAC range decreases, and the absolute error shrinks even more. Reducing the phase range by 4×, from 45° to 11.25°, suppresses the absolute error by ~70×.

To test its effectiveness, the cosine-nonlinearity correction is tested on simulated WB sensor outputs (Fig. 3.11). For simplicity, an in-batch spread of ±5% on $R_S(T_0)\cdot C_0$ is assumed, with zero no TC spread on both the silicided sensing resistor or the MIM capacitor. To facilitate operation in the presence of process and large temperature variations, the phase range is set to ±22.5°.

Without any pre-processing, a linear fit on the modulator's bitstream average results in a systematic nonlinearity of ~6 °C over a 125 °C temperature range, which is mainly due to the nonlinear RC-to-phase characteristic of the WB (Fig. 3.12).

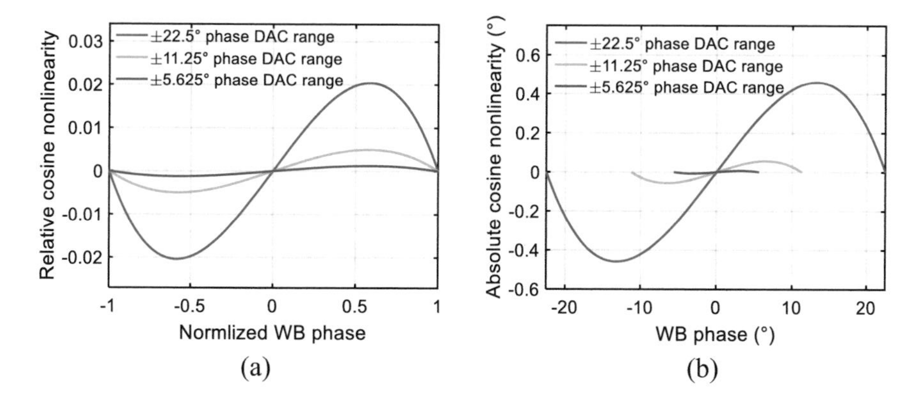

Fig. 3.10 (**a**) Relative and (**b**) absolute cosine nonlinearity with different phase DAC ranges

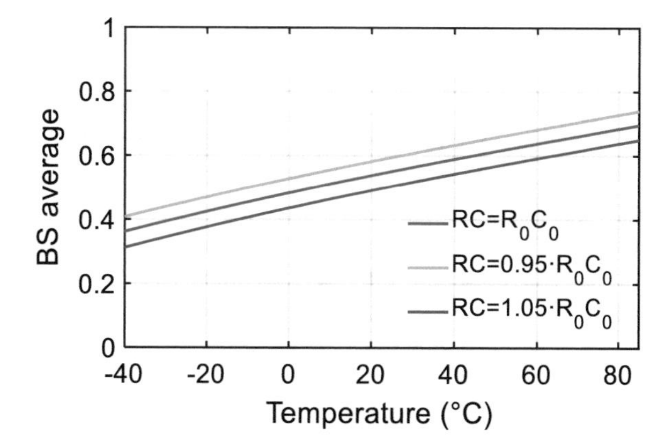

Fig. 3.11 Simulated WB sensor output with a phase DAC range of ±22.5°

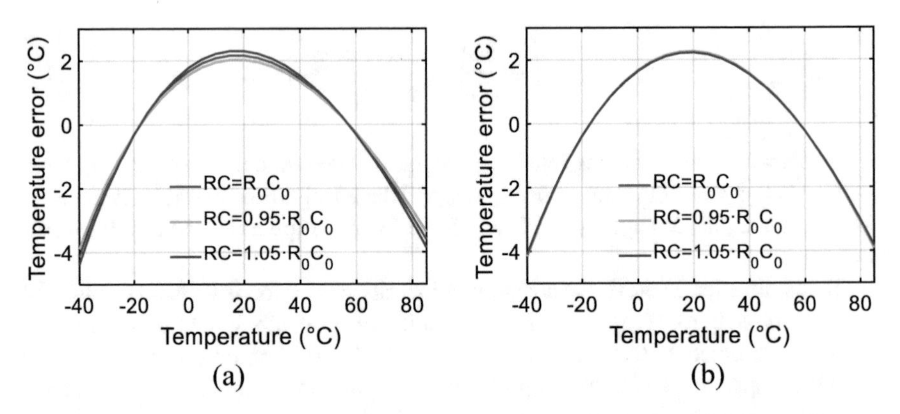

Fig. 3.12 Residual error of WB sensors after a linear fit assuming an ideal phase-domain ADC: (**a**) without cosine nonlinearity compensation and (**b**) with cosine linearity compensation

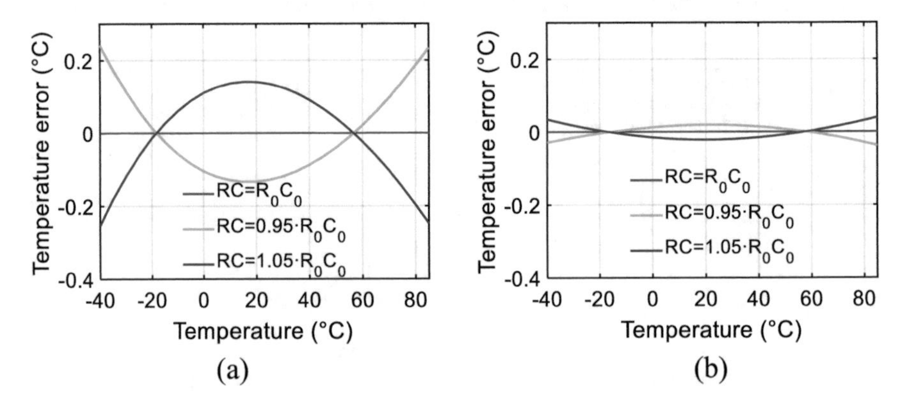

Fig. 3.13 Residual error after systematic error removal: (**a**) without cosine nonlinearity compensation and (**b**) with cosine linearity compensation

After removing this systematic nonlinearity with a fifth-order polynomial, the maximum residual error becomes 0.24 °C. After applying the cosine nonlinearity correction (Eq. 3.12), the systematic nonlinearity remains roughly the same, but the spread after a linear fit can be suppressed to 0.03 °C, as shown in Fig. 3.13.

Although the method is effective, there is still significant error left, which can be further suppressed by a more complete nonlinearity correction. Since Eq. (3.12) assumes the use of sine-wave excitation/demodulation, better results can be obtained by using a simulated μ to RC mapping (Fig. 3.14a) based on square-wave excitation/demodulation to eliminate all nonlinearity errors. After applying this mapping, represented by a seventh-order polynomial, the effect of resistor spread can be completely corrected by a linear fit, as shown in Fig. 3.14b.

It is worth mentioning that the resulting μ to RC nonlinearity, just like the cosine-nonlinearity, decreases with the phase DAC range. For example, with a ±5.625°

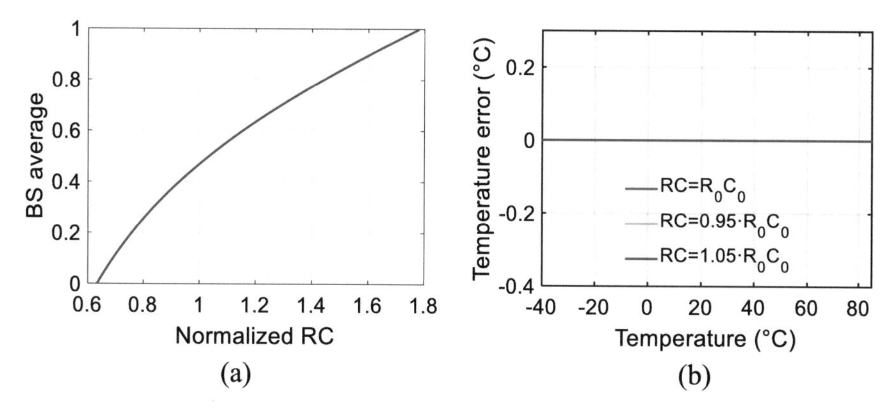

Fig. 3.14 (**a**) Normalized RC versus BS average of WB sensors with a phase DAC range of ±22.5°. (**b**) Residual error after applying the μ to R nonlinearity correction

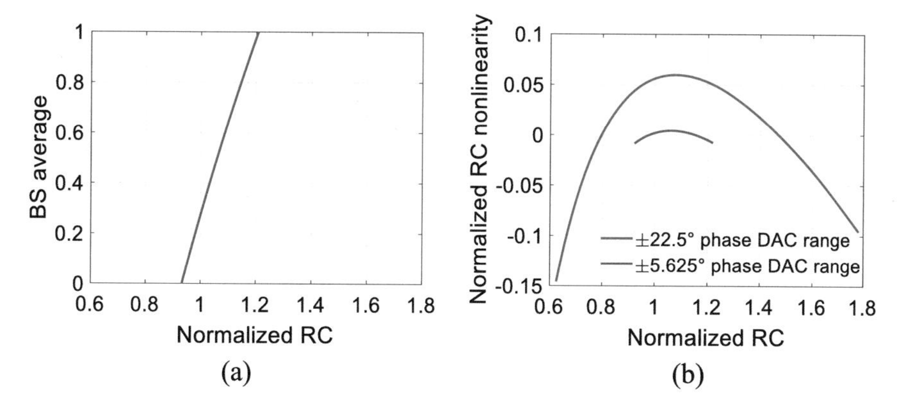

Fig. 3.15 (**a**) Normalized RC versus BS average of WB sensors with a phase DAC range of ±5.625° (**b**) RC nonlinearity compared with a ±22.5° phase DAC

DAC range, the absolute RC nonlinearity will become ~20× smaller (Fig. 3.15). As a result, a nonlinearity correction is not required to achieve decent inaccuracy after trimming.

3.3 Implementation I, Proof of Concept[1]

As a proof of concept, a WB sensor was built to investigate the potential energy efficiency and accuracy of resistor-based sensors. However, some other important parameters, for example, chip area, and 1/*f* noise, were not optimized.

[1] Pan et al. [16].

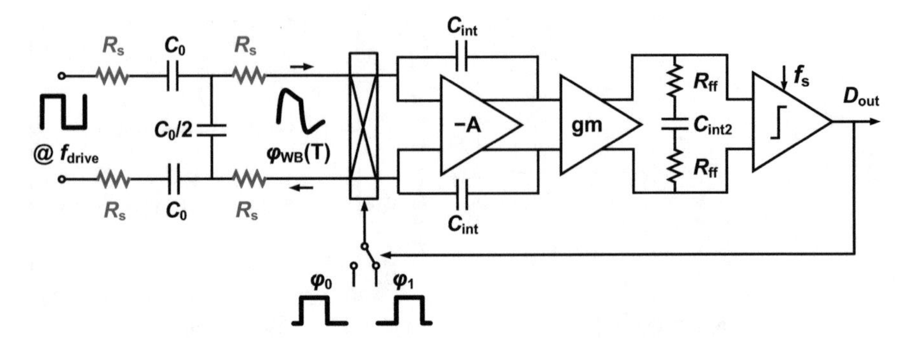

Fig. 3.16 Simplified circuit diagram of the first WB sensor implementation

3.3.1 Circuit Implementation

The block diagram of the proposed temperature sensor is shown in Fig. 3.16. For high resolution, the WB resistor (silicided poly, 1 μm wide) was chosen to be 32 kΩ, while the WB capacitor was set to 10 pF, resulting in a center frequency f_0 of 500 kHz. The filter is driven by a square-wave signal at this frequency, that is, $f_{drive} = f_0 = 500$ kHz. This is derived from an 8 MHz external master clock by a divide-by-16 circuit. For simplicity, the PDΔΣ-ADC employs a single-bit quantizer. Its two square-wave phase references, $\varphi_0 = 67.5°$ and $\varphi_1 = 112.5°$, are also generated by the divider circuitry. Their phase difference of 45° is chosen to accommodate the spread of the WB's resistors and capacitors, and hence in $\varphi_{WB}(T)$.

In contrast to a previous design, which was based on a first-order modulator [1], this design employs a second-order modulator to achieve sub-mK resolution in a short (5 ms) conversion time. As shown in Fig. 3.16, it employs a feed-forward topology [4], which requires only one feedback DAC, and also reduces the swing in the loop filter. To establish a low-impedance virtual ground at the input of the ADC, the first stage consists of an active integrator, while, for simplicity, the second stage consists of a gm-C integrator. The feedforward coefficient is realized by the introduction of R_{ff} in series with the integration capacitor (C_{int2}) of the second stage. Its output is sampled at f_{drive} by a comparator, which is triggered at $\varphi_{trig} = 135°$.

3.3.1.1 Chopper and Chopper Merging

To suppress its $1/f$ noise, the opamp of the first stage is chopped. By making the chopping frequency f_{chop} the same as the driving frequency f_{drive}, the input chopper and the input demodulator can be merged into a single chopper in series with the integration capacitors, as shown in Fig. 3.17. This chopper merging technique [5] simplifies the required control logic and minimizes errors due to charge injection mismatch.

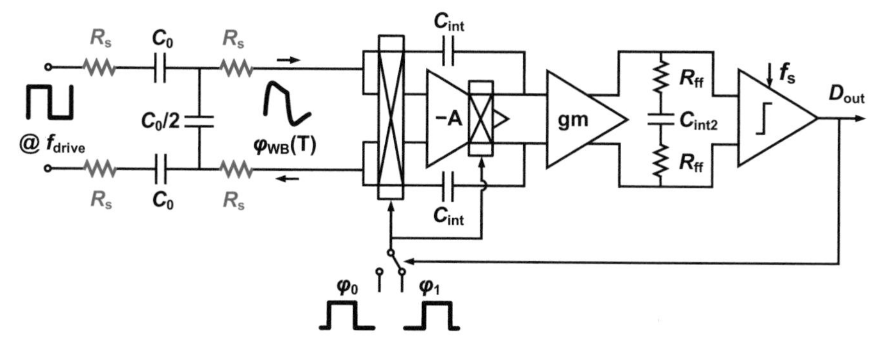

Fig. 3.17 Simplified circuit diagram of the first WB sensor implementation with a merged chopper

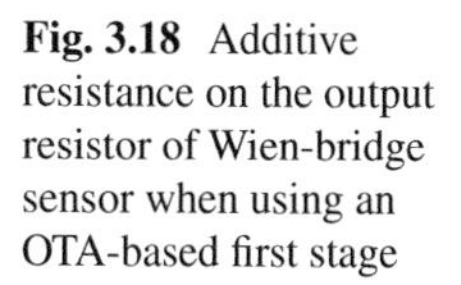

Fig. 3.18 Additive resistance on the output resistor of Wien-bridge sensor when using an OTA-based first stage

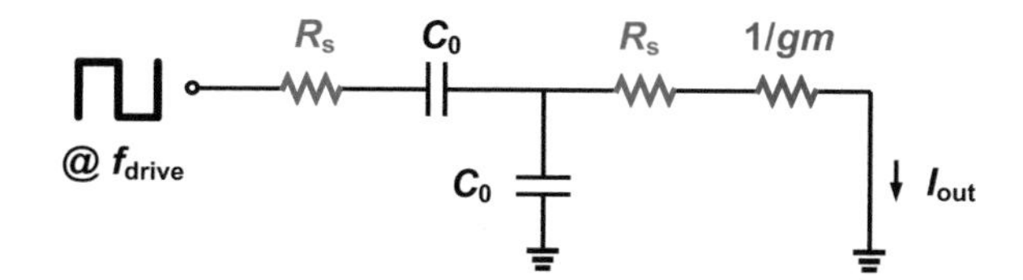

3.3.1.2 Amplifier Design

In principle, the first-stage amplifier can be implemented as an energy-efficient single-stage operational trans-conductance amplifier (OTA). However, the input impedance of the resulting integrator is then approximately 1/gm, where gm is the OTA's transconductance. As shown in Fig. 3.18, this resistance loads the WB, thus altering $\varphi_{WB}(T)$ and degrading its temperature-sensing accuracy. For example, with the chosen s-p-poly resistors, a 10% variation on a nominal gm of 1 mS (I_d ~ 50 μA, or 4× larger than the maximum output current of the WB) will translate into a temperature-sensing error of more than 0.5 °C.

In a two-stage amplifier, the input stage does not need to provide the output current. This helps to reduce its input swing and hence the input impedance of the first integrator, which in turn results in a smaller temperature-sensing error. For simplicity, and to avoid the need for Miller compensation capacitors, the gain of the output stage should not be large, so that the pole formed by the gm/C_{load} of the output stage, is well beyond the unity-gain frequency of the amplifier. In this work, a two-stage opamp consisting of a telescopic gm stage followed by two low-V_T PMOS source followers was used, as shown in Fig. 3.19. The common-mode feedback of the gm stage is realized by two PMOS transistors in their triode region. The tail current of the telescopic gm, which is 16 μA at room temperature, is optimized for low noise and power consumption. The source followers' bias current (20 μA/branch at room temperature) is chosen to handle the WB's peak current (11 μA at room temperature, 16 μA at −40 °C). The opamp's 1/*f* noise has a corner frequency of about 15 kHz, and so is effectively canceled by chopping at 500 kHz. To limit the first stage's output swing (which includes chopper ripple) and thus to relax the design of

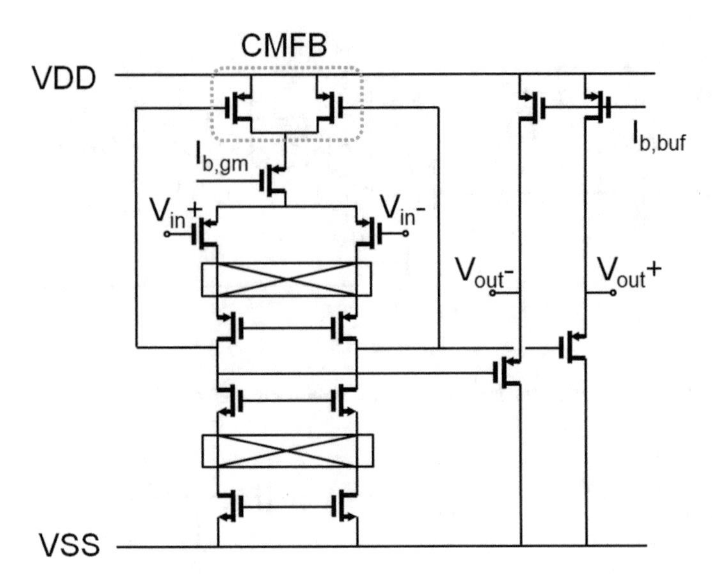

Fig. 3.19 Schematic of the opamp in the first stage

the gm-C second stage, the integration capacitor of the first stage is made quite large (180 pF each).

The gm-C second stage is built around a telescopic OTA with source degenerated input pairs, which achieves a good balance between energy efficiency and linearity. It draws 4 μA, which is less than 10% of that of the first stage.

3.3.2 Measurement Results

The sensor was fabricated in a standard 0.18 μm CMOS technology, and the chip micrograph is shown in Fig. 3.20. For flexibility, a sinc^2 decimation filter is implemented off-chip. Each sample contains two different temperature sensors: one with silicided p-poly (s-p-poly) resistors, and for the sake of comparison, the other with nonsilicided n-poly resistors. These two co-integrated sensors share the same constant-gm biasing and phase generation circuits. Each sensor occupies an active die area of 0.72 mm^2, about 40% of which is consumed by the first integrator's capacitors (2 × 180 pF). Each sensor draws 87 μA from a 1.8 V power supply, including the readout circuits. At room temperature, DC supply sensitivities of −0.17 °C/V (s-p-poly bridge) and 0.34 °C/V (n-poly bridge) were observed for supply voltages ranging from 1.6 to 2 V.

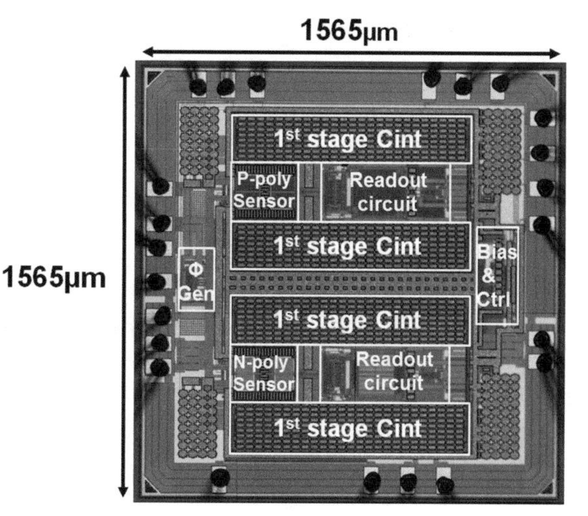

Fig. 3.20 Micrograph of the first WB sensor implementation

3.3.2.1 Resolution and FoM

Since the phase output of the WB sensor is determined by its driving frequency, random jitter will translate into random phase noise and degrade the sensor's resolution. To prevent this, the sensors are driven by a low-jitter (1 ps_{rms}) frequency reference, which only degrades the sensor's resolution by about 0.5%. Furthermore, the temperature of the sensors was stabilized by mounting them inside a cavity in a large (10 kg) metal block, which, in turn, was placed in a temperature-controlled oven (Vötsch VT7004). More details of the setup can be found in Appendix A.1.

The power spectral densities of both sensors' output bitstreams are shown in Fig. 3.21, where 0 dB corresponds to 1 in the BS average amplitude, which full range is [−1, 1]. The sensor's noise floor is dominated by the RC-filters' thermal noise. The n-poly resistor exhibits a 1/*f* corner of about 10 Hz, while that of the s-p-poly sensor is below 1 Hz. Since the two sensors are readout in exactly the same way, the 1/*f* noise of the n-poly sensor can be directly attributed to the sensing resistors.

After decimating their bitstreams at room temperature (RT ~25 °C), the sensors' resolution is plotted versus conversion time, as shown in Fig. 3.22. To suppress the effects of ambient temperature drift, the resolution was determined from a two-sample Allan deviation, that is, from the difference between two successive measurements. In a 5 ms conversion time (2500 samples), the calculated resolutions of s-p-poly and the n-poly sensors are 410 μK_{rms} and 880 μK_{rms}, respectively.

However, the differencing operation inherent to the Allan deviation calculation will also suppress the sensor's non-negligible 1/*f* noise (Fig. 3.22). To avoid this, the standard deviation can be computed over a shorter interval (1 second), during which the temperature drift will be negligible compared to the sensor's noise. Given the same 5 ms conversion time, the resolution becomes 440 μK_{rms} for the s-p-poly

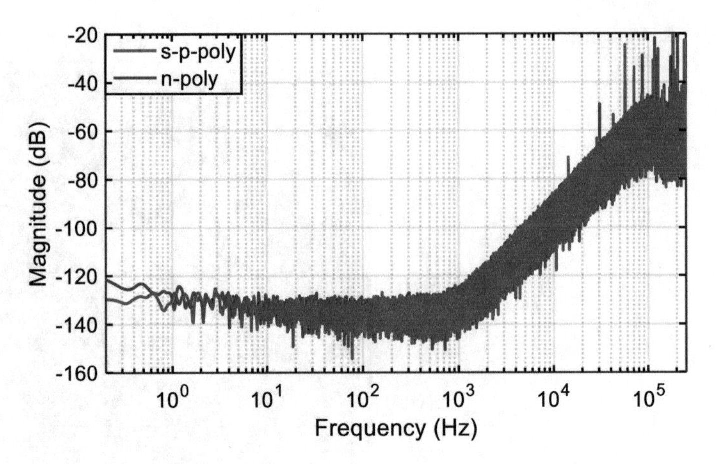

Fig. 3.21 The power spectral densities of the output bitstream of both sensors

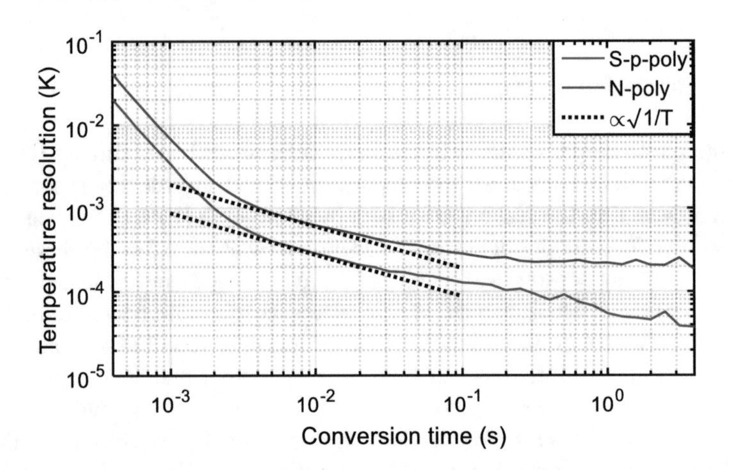

Fig. 3.22 The calculated resolution using a two-sample Allan deviation

sensor and 990 μK_{rms} for the n-poly sensor. The corresponding resolution FoM of the s-p-poly sensor is then 0.15 pJ·K^2.

3.3.2.2 Calibration and Inaccuracy

Twenty samples from one wafer in ceramic DIL packages were characterized from −45 °C to 85 °C (steps of 10 °C) in a temperature-controlled oven. The actual temperature was established by a calibrated Pt-100 RTD. To partially compensate for the spread of f_0 with process, f_{drive} was set to 562.5 kHz (9 MHz master clock) instead of the nominal 500 kHz. After the fixed nonlinearity correction introduced in Sect. 3.2.3.2, the extrapolated resistance versus temperature plots (R-T plots) of the

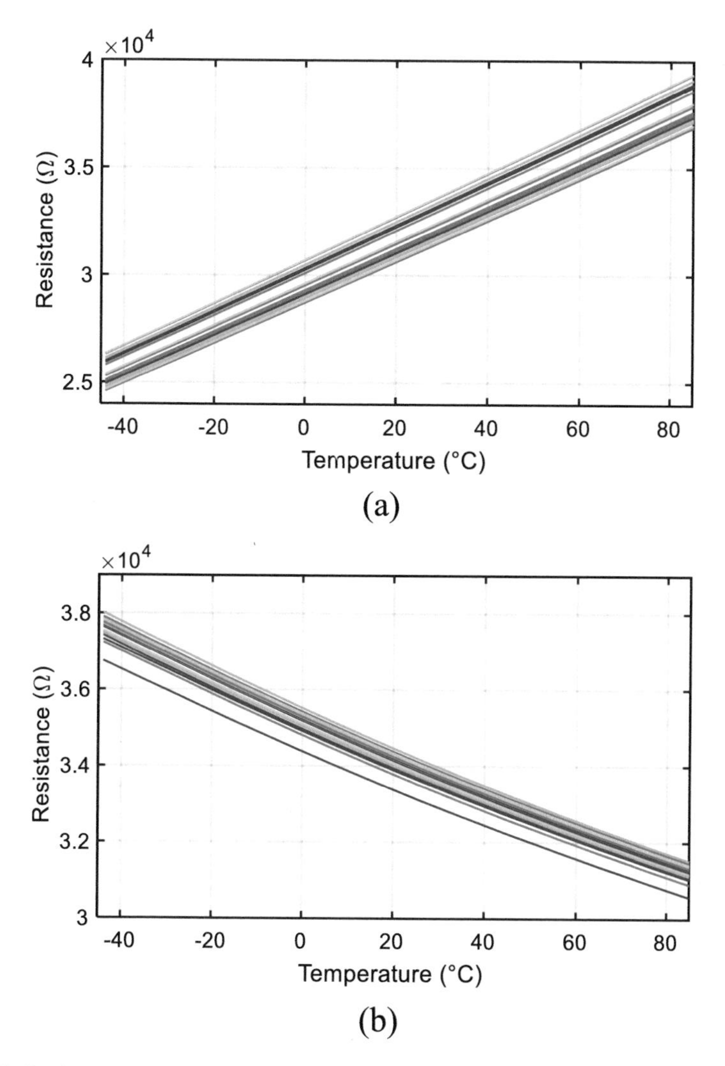

Fig. 3.23 Resistance versus temperature plots of (**a**) s-p-poly sensor (**b**) n-poly sensor in ceramic packages

s-p-poly and the n-poly sensors are shown in Fig. 3.23, and the corresponding average R-T plots are shown in Fig. 3.24. The corresponding 1st- and second-order TCs agree well, to within a few percent, with the models provided by the foundry, thus validating the nonlinearity correction technique. After a first-order linear fit, the remaining nonlinearity, mainly due to the nonlinearity of the sensing resistor's TC, is quite systematic (Fig. 3.25), and so can be removed by a fixed third-order polynomial obtained by batch calibration. After this systematic nonlinearity correction, the s-p-poly sensor achieves a 3σ inaccuracy of ±0.03 °C, while the n-poly sensor's

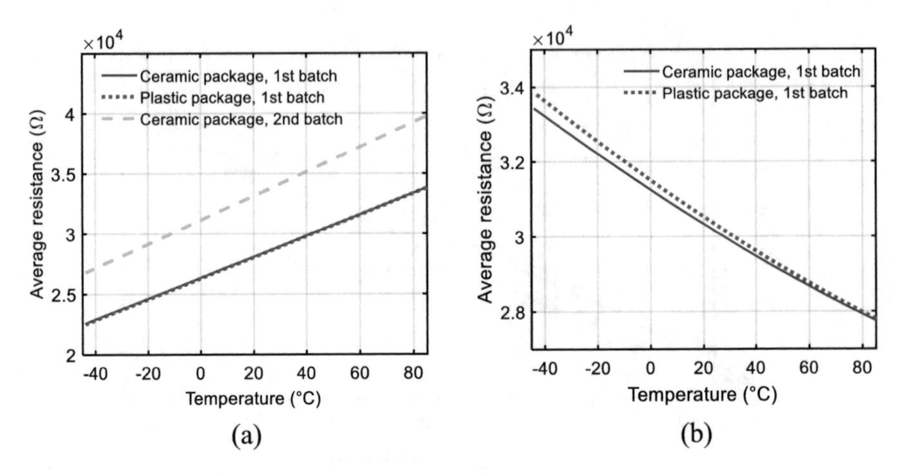

Fig. 3.24 Average resistance versus temperature of (**a**) s-p-poly sensor and (**b**) n-poly sensor

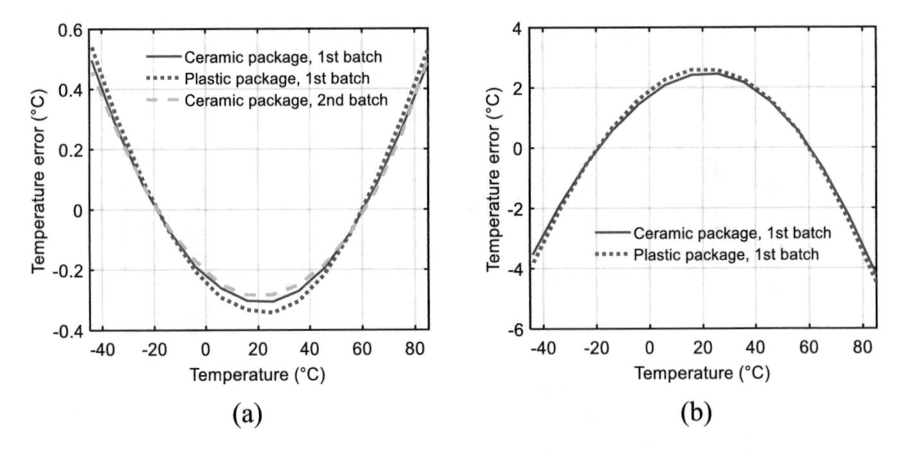

Fig. 3.25 Systematic temperature nonlinearity of (**a**) s-p-poly sensor (**b**) n-poly sensor after a first-order fit

inaccuracy is about ±0.3 °C, as shown in Fig. 3.26. Without this correction, the accuracy of the s-p-poly sensor would have been about 2× worse.

To reduce calibration costs, the number of calibration temperatures should be reduced as much as possible. So rather than doing a first-order fit based on data obtained at multiple temperature points, a more straightforward two-point calibration can be done. This results in only a slight loss of accuracy: when calibrated at −15 °C and 65 °C, the s-p-poly sensor achieves a 3σ inaccuracy of ±0.05 °C.

Serendipitously, the TC and the RT resistance (R_0) of the s-p-poly sensor were found to be correlated, as shown in Fig. 3.27a. By exploiting this correlation, a 3σ inaccuracy of ±0.2 °C could be achieved after a single-point calibration, as shown in Fig. 3.28. Unfortunately, this correlation is much weaker for n-poly sensor (Fig. 3.27b), and the 3σ inaccuracy after a single-point calibration is only ±0.6 °C.

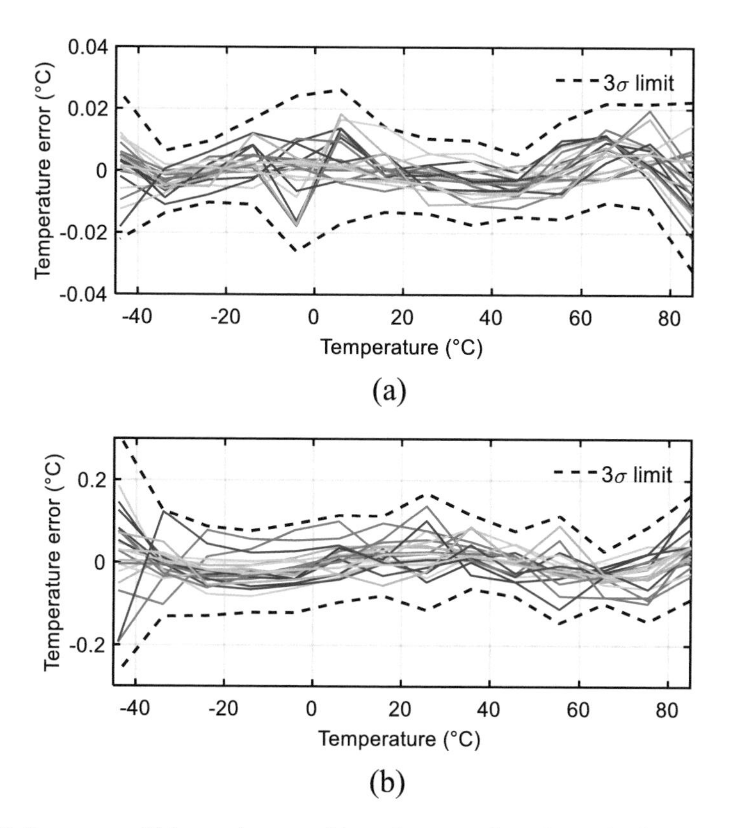

Fig. 3.26 Inaccuracy of (**a**) s-p-poly sensor (**b**) n-poly sensor after a first-order fit and systematic nonlinearity removal

3.3.2.3 Plastic Packaging

In production, low-cost plastic packages are preferred over ceramic packages. However, the accompanying mechanical stress [6] impacts the sensor's accuracy. Because of the metal-like properties of silicided poly resistors, their stress sensitivity is much less than that of nonsilicided poly resistors. The average resistance versus temperature plot of 12 sensors produced in the same batch is shown in Fig. 3.24. Compared to the ceramic packaged chips, both the TC and R_0 of the n-poly resistors in plastic packages change significantly, while those of the s-p-poly resistors do not.

However, compared to ceramic packaged devices, a shift was observed in the TC-R_0 correlation of the s-p-poly sensors. Even based on the limited number of samples, the correlation also appears to be weaker, as shown by an outlier in Fig. 3.27a. After a two-point calibration, however, the change in their systematic nonlinearity is less than 0.05 °C (Fig. 3.25). After a packaging-specific systematic nonlinearity correction, the sensor achieves a 3σ inaccuracy of ±0.2 °C, as shown in Fig. 3.29, mainly due to the outlier.

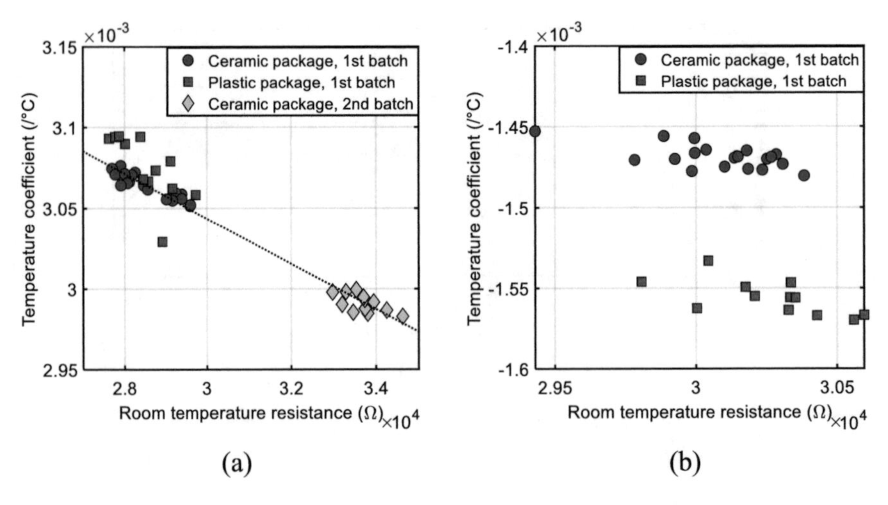

Fig. 3.27 Correlation between R_0 and its TC for (**a**) the s-p-poly resistor and (**b**) the n-poly resistor

3.3.2.4 Batch-to-Batch Spread

To verify the effect of batch-to-batch spread on the s-p-poly sensor's inaccuracy, 12 devices from a different batch (fabricated a few months after the first batch) were characterized in ceramic packages. As shown in Fig. 3.24, however, the center frequency f_0, and hence the extrapolated resistance of the s-p-poly sensors then shifted by about 16%. This resistance shift was compensated by reducing f_{drive} by 16% during characterization. The sensor's extrapolated TC-R_0 relationship is shown in Fig. 3.27. Despite the significant shift in f_0, the linear correlation discussed in Sect. 3.3.2.2 is still valid. The s-p-poly sensor achieves an estimated 3σ inaccuracy of ±0.3 °C after a correlation-assisted single-point calibration. After an individual first-order fit, the maximum difference in the systematic nonlinearity of the two batches is 0.04 °C from −40 °C to 85 °C (Fig. 3.25).

3.3.2.5 Comparison with Prior Art

The performance of the s-p-poly sensor is summarized in Table 3.1 and compared with that of earlier energy-efficient CMOS temperature sensors. It achieves an energy efficiency of 0.15 pJ·K^2, which is over 4× better than an earlier resistor-based sensor [8]. When packaged in ceramic, the sensor achieves an inaccuracy of ±0.03 °C (3σ) after a first-order fit followed by a fixed systematic nonlinearity correction. It also achieves ±0.2 °C (3σ) after a single-point calibration, which is comparable to that of most BJT-based sensors [7, 11].

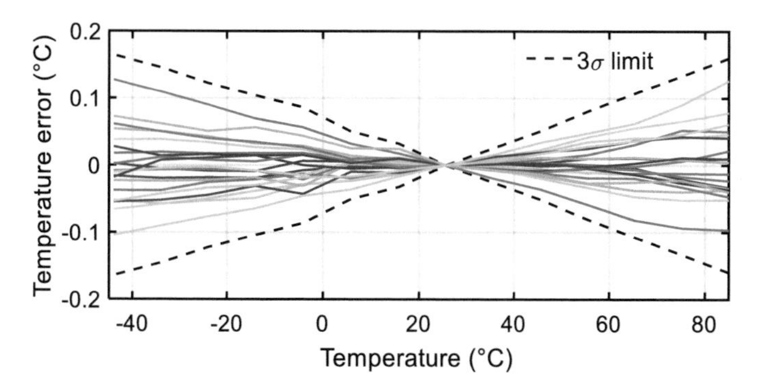

Fig. 3.28 Inaccuracy of s-p-poly sensor after a correlation-assisted 1-point calibration (ceramic packaged, first batch)

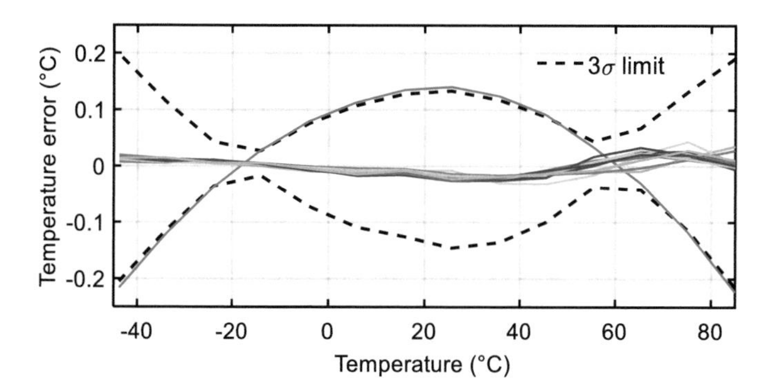

Fig. 3.29 Inaccuracy of plastic packaged s-p-poly sensor after a first-order fit and systematic nonlinearity removal

Table 3.1 Performance summary of the first sensor implementation and comparison with previous work

	ISSCC'14 [7]	JSSC'15 [8]	JSSC'15 [9]	Implementation I
Sensor type	BJT	Resistor WhB	Resistor WB	Resistor WB
Technology	0.7 μm	0.18 μm	0.18 μm	0.18 μm
Area [mm²]	1.5	0.43	0.09	0.72
Temp. Range [°C]	−40—130	−40—125	−40—85	−40—85
3σ Inaccuracy [°C] (trimming points)	0.3 (1)	0.4[a] (2[b])	0.12[a] (3)	0.03 (2[c])
Relative inaccuracy	0.35%	0.48%	0.19%	0.05%
Power [μW]	160	65	31	160
Conv. time [ms]	1.8	0.1	32	5
Resolution [mK]	3	10	2.8	0.44
Res. FoM[fJ·K²]	3200	650	8000	150

[a]Min/Max. [b]1-point trim with a fixed 1st-order fit. [c]1st-order fit

3.4 Implementation II, Reduced Chip Area[2]

Although the first implementation achieves good performance, it has a large chip area (0.72 mm^2) and a limited temperature range. This section describes a compact (0.12 mm^2) resistor-based sensor that uses a scaled WB filter to achieve 0.1 °C (3σ) inaccuracy over a wider temperature range of 220 °C.

3.4.1 Circuit Implementation

As shown in Sect. 3.3, most of the chip area in the first implementation is consumed by the integration capacitor in the first stage (C_{int}). Since $C = Q/U = I \cdot T/U$, there are three options: reducing the WB input current, shortening its period (i.e., increasing f_{drive}), or increasing the output swing of the first integrator.

In this design, the WB resistor width is halved (1 µm to 0.5 µm) compared to that in the first implementation, leading to a doubled resistance (64 kΩ) at the expense of worse resolution. The designed WB driving frequency f_{drive} is kept the same (500 kHz). This is because the temperature information is embedded in the shape of the input current, and the accuracy will be degraded if the bandwidth of the readout circuit is not sufficiently higher compared to f_{drive}. As a result, the WB capacitor is also halved (5 pF instead of 10 pF) to accommodate the change of the WB resistance.

Increasing the first integrator's output swing requires a new opamp topology. The first integrator is based on a two-stage Miller-compensated opamp based on current-reuse amplifiers, as shown in Fig. 3.30. The first stage provides good energy efficiency, while the second uses high-V_T devices to efficiently provide a nearly rail-to-rail output swing. Compared to the conventional choice of two common-source stages, it provides twice the output current for the same bias current. Together with the doubling of R_S (e.g., halved WB current), the enlarged opamp output swing allows the value of C_{int} to be reduced from the 180 pF used in the first implementation, to 23 pF. At room temperature (RT), the amplifier draws 14 µA, achieves 80 dB gain, and has a gain-bandwidth product of 17 MHz with a 500 fF loading capacitor.

To further reduce area, the second integrator and the feed-forward coefficient are realized in a switched-capacitor manner, thus avoiding the large resistors used in the first implementation, as shown in Fig. 3.31. The associated folded-cascode amplifier draws only 2.5 µA at RT. The sampling capacitor C_{S2} and the feedforward capacitor C_{ff2} are 100 fF and 200 fF, respectively, while the second-stage integration capacitor C_{int2} is 1 pF.

The phase references $\varphi_{0,1}$ are 90° ± 30° instead of 90° ± 22.5° to extend the measurable temperature range. Like f_{drive} and $\varphi_{a,b}$, they are derived from an external 6 MHz frequency reference using on-chip logic.

[2] Pan et al. [17].

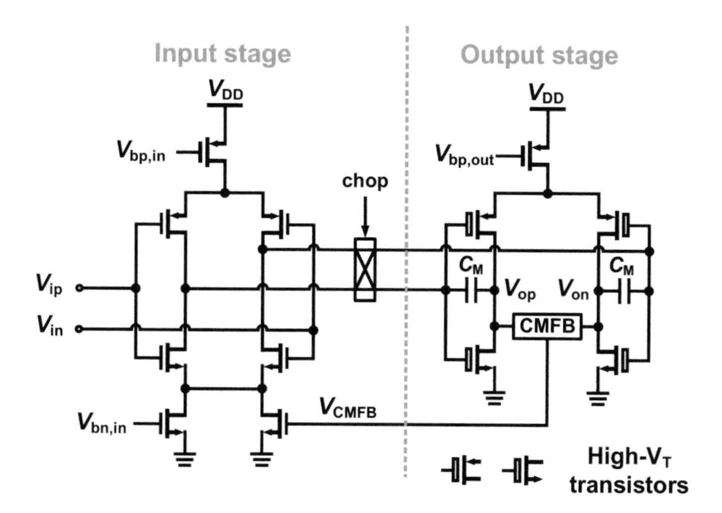

Fig. 3.30 Simplified diagram of the first-stage amplifier

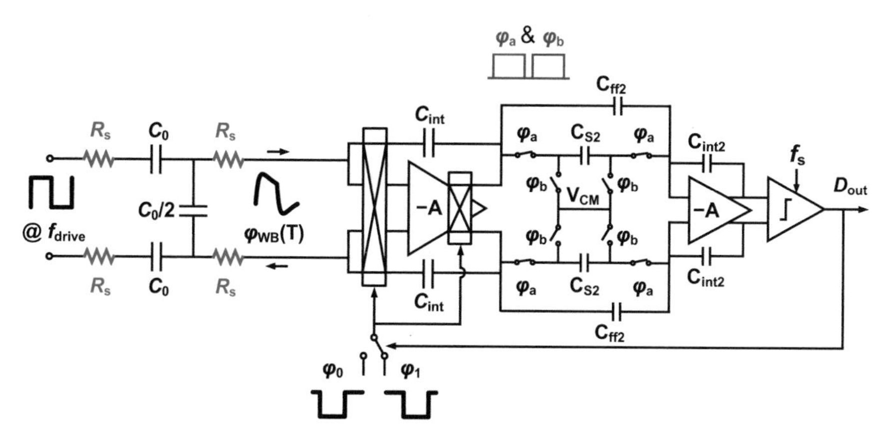

Fig. 3.31 Simplified circuit diagram of the second implementation

However, reducing C_{int} will increase the opamp's closed-loop input impedance Z_{in} ($\propto 1/(C_{\text{int}}$*GBW)), where GBW is the opamp's gain-bandwidth product. This is in series with the WB (Fig. 3.18) and is thus a source of spread and $1/f$ noise. To minimize spread, a constant-Gm biasing circuit based on the same resistor type as R_S ensures that Z_{in} tracks R_S over a wide temperature range. Although the opamp is effectively chopped, the bias current's $1/f$ noise will modulate Z_{in}, and thus R_S, causing residual $1/f$ noise. To minimize this noise, the core of the biasing circuit was realized with large PMOS devices (W/L = 40 μm/5.5 μm), and critical current mirrors were realized with the standard NPN transistors available in the chosen process (Fig. 3.32). Simulations show that the sensor's $1/f$ corner is then about 1 Hz and that R_S is less than 1% Z_{in} over corners.

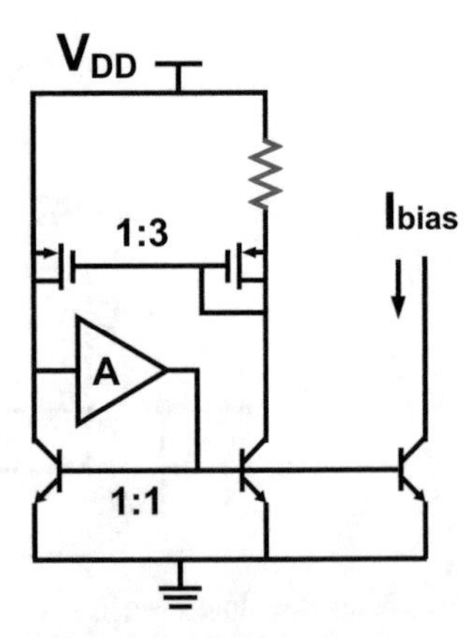

Fig. 3.32 Biasing circuit

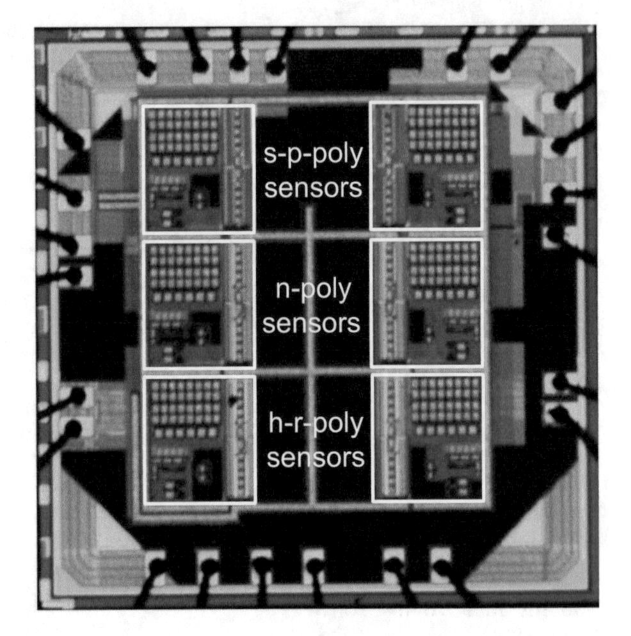

Fig. 3.33 Micrograph of the second WB sensor implementation

3.4.2 Measurement Results

Three pairs of identical sensors based on silicided p-poly (s-p-poly), unsilicided n-poly (n-poly), and high-resistive poly (h-r-poly) resistors were fabricated on the same die in a 0.18 μm CMOS process (Fig. 3.33). This facilitates the use of differential measurements to reject temperature drift of the ambient environment, so that the sensor resolution can be accurately measured. Each sensor consumes 29 μA from a single 1.8 V supply and occupies 0.12 mm^2, of which the WB occupies 25%.

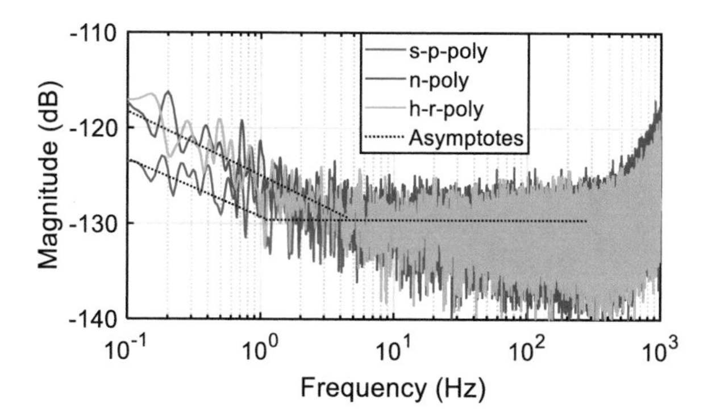

Fig. 3.34 Bitstream spectra (100 s interval, Hanning window) of different WB sensors

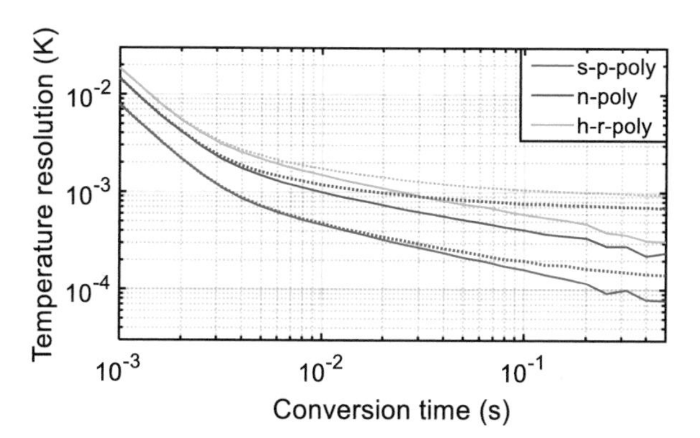

Fig. 3.35 Resolution versus conversion time of different WB sensors given a 1 s (solid lines, 100× averaging) and 100 s (dashed lines) time interval

3.4.2.1 Resolution and FoM

Bitstream spectra of the three sensors are shown in Fig. 3.34 based on differential data captured from 100 s time. The corner frequencies are ~4 Hz (n-poly and h-r-poly) and ~1 Hz (s-p-poly). The latter is limited by the readout electronics. Sensor resolution is derived from the standard deviation within a 1 s time interval (Fig. 3.35). Due to its greater TC, the silicided-p-poly sensor exhibits the best resolution: 460 μK in a 10 ms conversion time, corresponding to a 110 fJ·K^2 resolution FoM. With a longer interval (100 s), however, the integrated 1/*f* noise power gets larger, resulting in a worse calculated resolution.

3.4.2.2 Calibration and Inaccuracy

A total of 10 chips (60 sensors) from a single batch were packaged in ceramic DIL and characterized from −40 °C to 180 °C. To correct for the inherent cosine nonlinearity of the PDΔΣM and the nonlinear relationship between φ_{WB} and R_S (Sect. 3.2.3.2), a seventh-order polynomial is used to translate the decimated output of each sensor into an equivalent sensor resistance R_S. Since the temperature dependency of polysilicon resistors is comparatively linear, this approach minimizes the

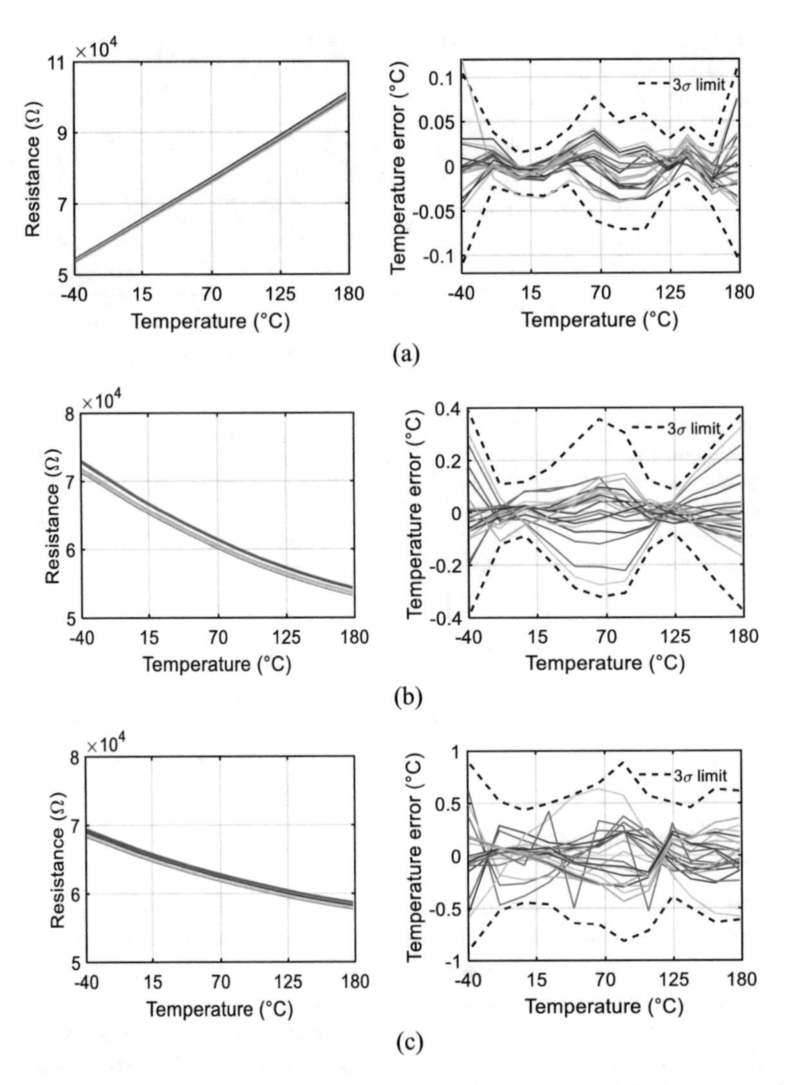

Fig. 3.36 Extracted sensor resistance R_S of ceramic packaged sensors (left); inaccuracy after a first-order fit and systematic nonlinearity correction (right) of (**a**) s-p-poly (**b**) n-poly and (**c**) h-r-poly sensors

residual error after a first-order fit. Figure 3.36 (left) shows the resulting temperature dependence of each resistor type. The following RT TCs were extracted: 0.31%/°C (s-p-poly), −0.15%/°C (n-poly), and −0.10%/°C (h-r-poly), which agree with the process documentation. After a first-order fit to compensate for process spread, followed by a fixed sixth-order polynomial to correct for systematic nonlinearity, the sensors' residual spread is shown in Fig. 3.36 (right). The sensors achieve 3σ inaccuracies of 0.1 °C (s-p-poly), 0.4 °C (n-poly), and 0.9 °C (h-r-poly) from −40 °C to 180 °C.

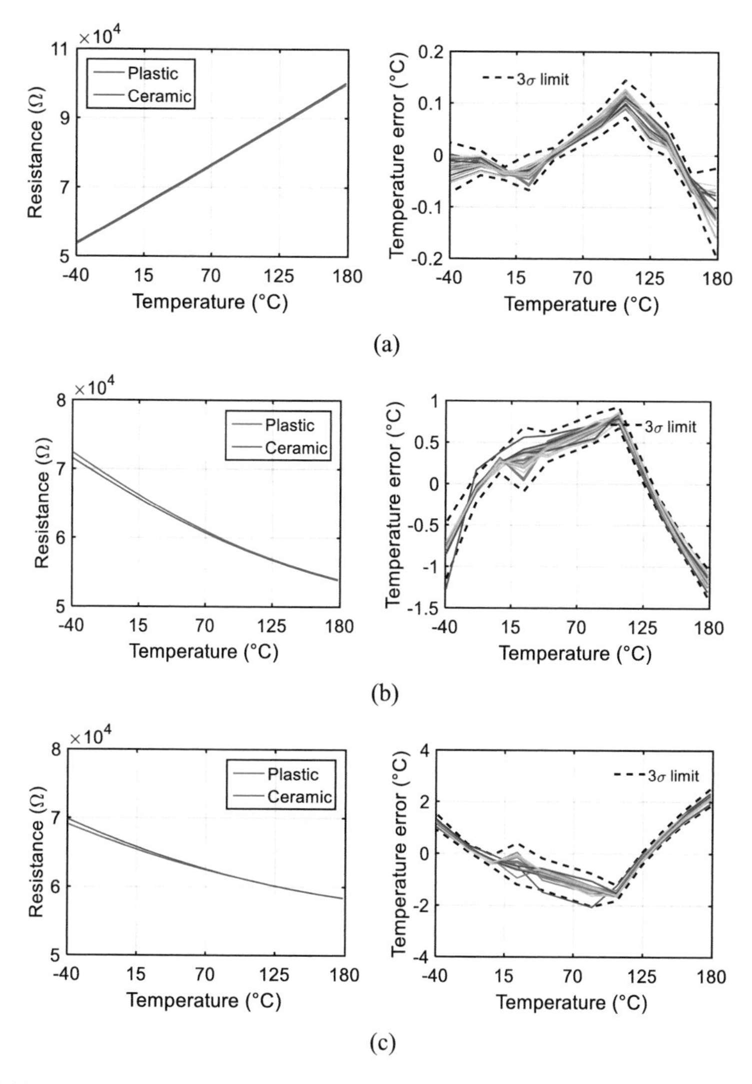

Fig. 3.37 Average resistance versus temperature of three types of resistors in different packages (left); inaccuracy of plastic packaged sensors after a first-order fit and system nonlinearity correction from ceramic packaged sensors (right) of (**a**) s-p-poly (**b**) n-poly and (**c**) h-r-poly sensors

To observe the effects of mechanical stress, 10 chips from the same batch were characterized in injection-molded plastic QFN packages. Figure 3.37 (left) shows the average dependency of R_S for both the ceramic- and plastic-packaged chips. After using the same nonlinearity correction polynomials determined for the ceramic-packaged chips, the inaccuracy after a first-order fit increases by only 0.2 °C for the s-p-poly sensors, but to 1.4 °C for the n-poly sensors, and even 2.5 °C for the h-r-poly sensors (Fig. 3.37 (right)). The sharp inflexion in all the inaccuracy plots around 100 °C is probably due to the effects of moisture on the plastic packages [10]. Of the three resistor types, the s-p-poly resistor is clearly the least stress-sensitive, exhibiting a packaging shift similar to that of BJT-based sensors [11].

3.4.2.3 Comparison to Implementation I

Compared to the first implementation, this design has a 76% larger operating range and occupies 6× less area, at the expense of a somewhat worse relative inaccuracy. Also, the sensor's 1/*f* noise corner is still limited by the readout circuit. These problems will be addressed in the third implementation.

3.5 Implementation III, Better Accuracy and Stability[3]

3.5.1 Circuit Implementation

To provide a fair comparison, both the WB sensor (64 kΩ s-p-poly resistor, 5 pF capacitor) and its driving frequency (500 kHz) in this third implementation are inherited from the second one. The phase DAC range is changed back to ±22.5°, as the sensor is now targeting the slightly narrower military temperature range from −55 °C to 125 °C.

As shown in Fig. 3.18, one primary error source of a WB sensor is the input impedance of the first-stage integrator, which is connected in series with the WB resistor. To minimize this Z_{in} ($\propto 1/(C_{int}{\cdot}GBW)$) without enlarging the chip area, the GBW of the first-stage opamp needs to be extended. However, this should not be at the expense of significantly increased opamp power.

In the second implementation, the opamp's GBW is limited by the current of the output stage and its load capacitor (C_L). For a sufficient phase margin, the frequency of the secondary pole (gm_2/C_L, where gm_2 is the transconductance of the opamp's output stage) should be 3× larger than that of the dominant pole (≈GBW), that is, GBW < 3 gm_2/C_L. The load capacitor (C_L) consists of the parasitic capacitance of the integrator ($C_{int,par}$), the drain capacitance of the output stage (C_d), the input capacitance seen from the CMFB circuit (C_{CMFB}), and that of the SC-based second stage (C_{2nd}).

The first two are hard to suppress. However, the effect of both C_{CMFB} and C_{2nd} can be removed by introducing a low-power buffer, as shown in Fig. 3.38, without changing the basic opamp structure (Fig. 3.30). Since the second stage of the ΔΣ-ADC only processes a DC signal, the speed requirement of the buffer is not stringent. In this work, the buffer is simply realized with PMOS source-followers, which only adds ~1.1 μA on the designed 16 μA baseline of the opamp. However,

[3] Pan et al. [18].

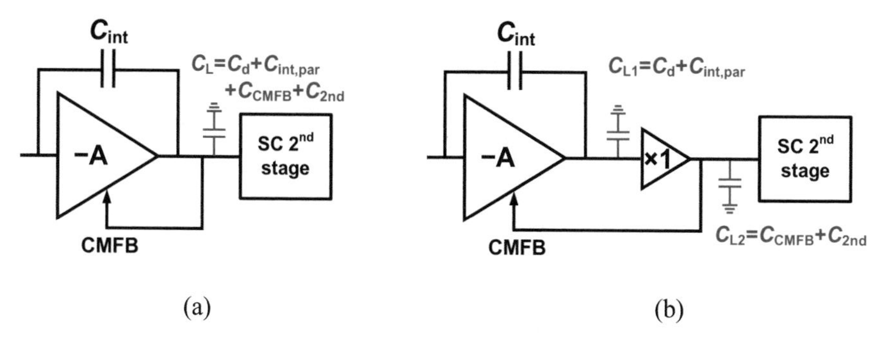

Fig. 3.38 Opamp connections with (**a**) all loading capacitors and (**b**) reduced loading capacitors

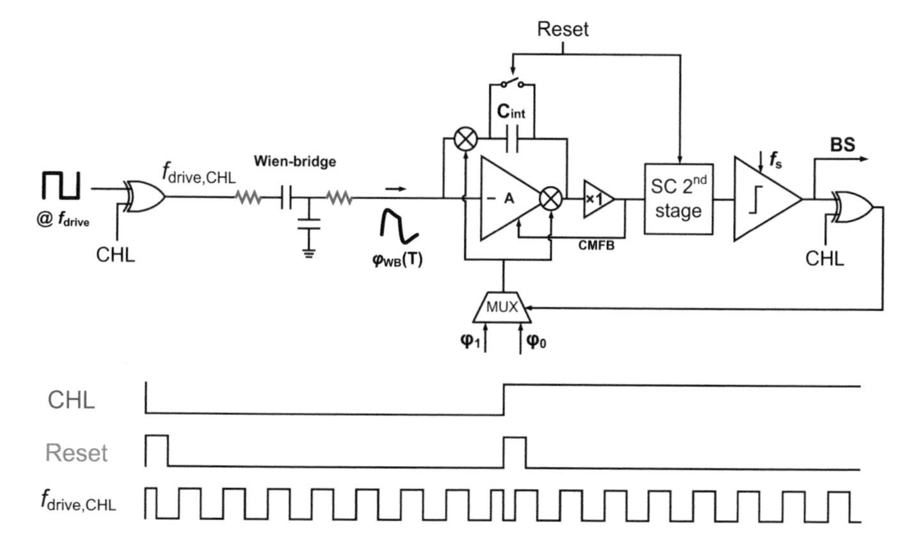

Fig. 3.39 Simplified circuit schematic and timing diagram of the sensor's low-frequency system-level chopping

after adjusting the Miller capacitor value accordingly, the opamp's GBW is extended from ~20 MHz to ~40 MHz, and thus the input impedance is suppressed by 2×.

This halved Z_{in} also reduces the $1/f$ noise coming from the biasing circuit. However, since only the input stage of the first-stage opamp is chopped (Fig. 3.30), the $1/f$ noise of the opamp output stage and the rest of the circuit remains. To further suppress this residual noise, XOR-gate-based low-frequency system-level choppers [12] are inserted to flip the entire system at 100 Hz. The driving signal (f_{drive}) is switched in the middle of its half period to minimize the settling error after this low-frequency chopping, as shown in Fig. 3.39. To avoid quantization noise folding, the ΔΣ-ADC works in an incremental mode, in which the integrators are reset at twice the system-level chopping frequency.

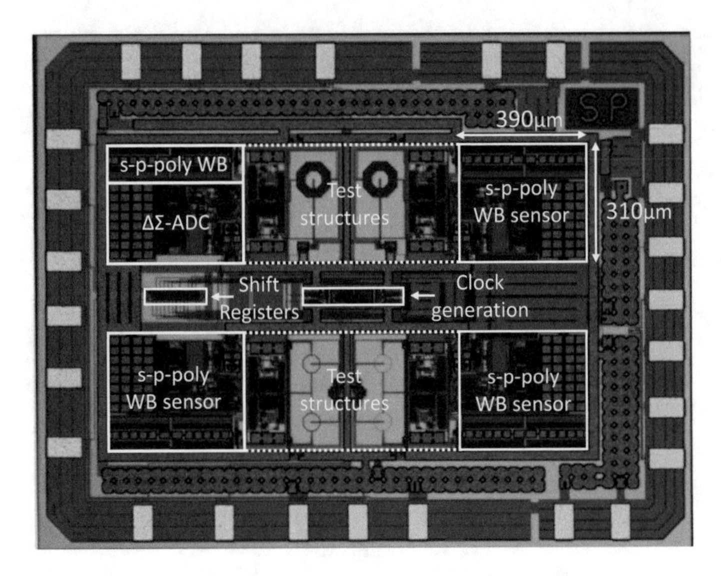

Fig. 3.40 Micrograph of the third WB sensor implementation

3.5.2 *Measurement Results*

With the same standard 0.18 μm process, four sensors based on s-p-poly resistors were fabricated on the same chip. Each sensor consumes 37 μA under a 1.8 V supply, and occupies 0.12 mm^2 after neglecting the test structures (Fig. 3.40), which is the same as the second implementation. For flexibility, the decimation filters (sinc^2) are implemented off-chip.

3.5.2.1 Resolution and FoM

Twenty samples from one wafer were characterized in a temperature-controlled oven after ceramic DIL packaging and mounting in good thermal contact with a large metal block. To reject the ambient temperature drift, the s-p-poly WB sensor's resolution is derived by computing the standard deviation of the difference in the output of two sensors from the same die. Figure 3.41 shows the s-p-poly sensor's output spectrum after decimation (with a 5 ms sinc^2 window). After enabling the system-level chopping (100 Hz), the sensor's $1/f$ noise corner is reduced from ~1 Hz to below 10 mHz. As shown in Fig. 3.42, over a 1 s interval, the sensor achieves a 450 μKrms resolution in a 10 ms conversion time (T_{conv}), which corresponds to a 0.13 pJ·K^2 resolution FoM.

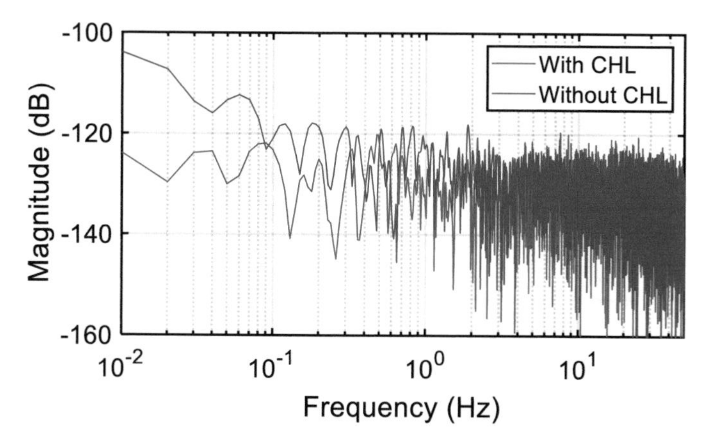

Fig. 3.41 Spectra of the sensor's decimated BS output with/without system-level chopping (CHL)

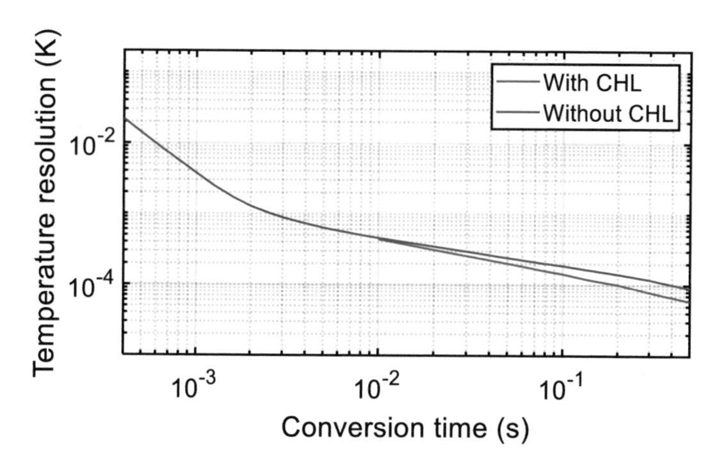

Fig. 3.42 Resolution versus conversion time over a 1 s period, with/without system-level chopping

3.5.2.2 Calibration and Inaccuracy

After converting the decimated ΔΣ-ADC output to WB resistance, an individual first-order fit is applied to remove process spread. Followed by a fifth-order systematic nonlinearity removal, the s-p-poly sensor achieves a 3σ inaccuracy of 0.03 °C from −55 °C to 125 °C, as shown in Fig. 3.43. A similar inaccuracy can be achieved after replacing the first-order fit with a two-point trim (−35 °C and 105 °C).

The correlation between the calculated WB resistance at RT (R_0) and its TC is shown in Fig. 3.44a. To fairly compare it with that in the first implementation, the correlation slope can be normalized by TC and R_0, that is,

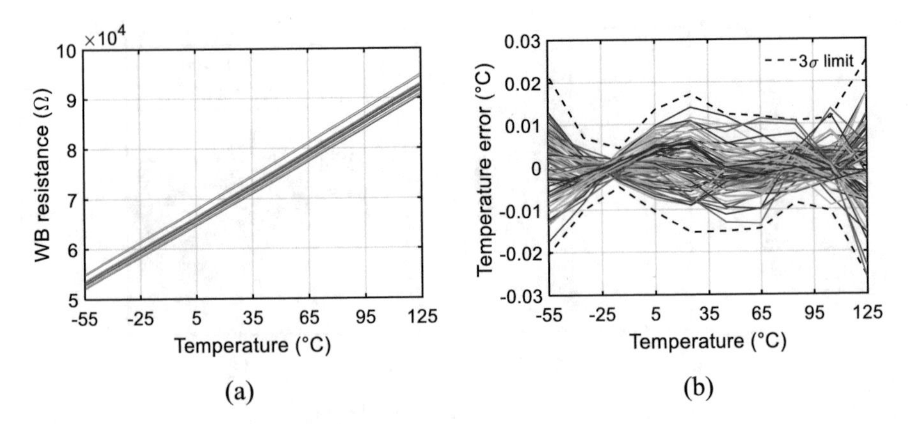

Fig. 3.43 (**a**) Extracted sensor resistance and (**b**) sensor inaccuracy after a first-order fit and systematic nonlinearity correction

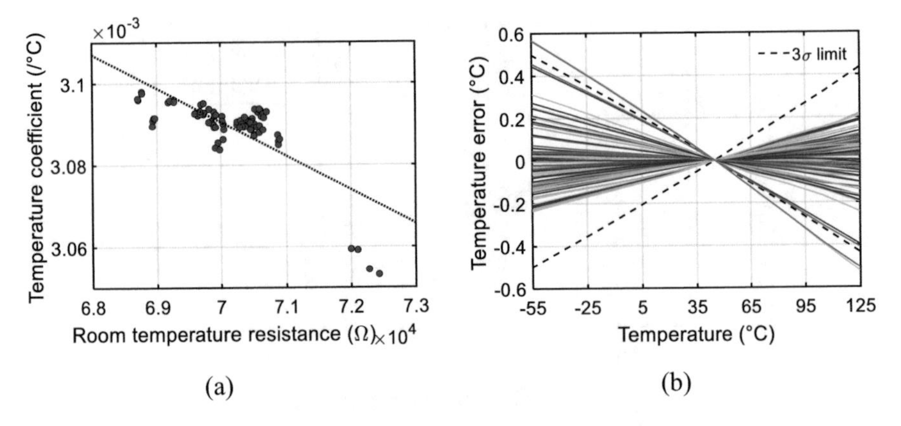

Fig. 3.44 (**a**) Correlation between R_0 and TC (**b**) sensor inaccuracy after a correlation-assisted 1-point trim and systematic nonlinearity correction

$$k_{\text{corr}} = \frac{dTC}{dR_0} \cdot \frac{R_0}{TC}. \tag{3.13}$$

In this design, k_{corr} is roughly ~−18%, which is larger than the −11% observed in the first WB implementation (Fig. 3.27a). However, the correlation is less significant. Consequently, the relative 1-point trimmed inaccuracy is degraded from 0.32% to 0.55%, and the 3σ error becomes 0.5 °C over the 1.4× larger temperature range, as shown in Fig. 3.44b. However, it is worth noticing that there are four outliers (from a single chip) that deviate from most of the samples. Without these outliers, the 1-point trimmed inaccuracy will become ~0.25 °C, which is similar to that achieved by the first WB implementation. The calculated k_{corr} under this situation is ~ − 7%, which is smaller than that of the first WB implementation.

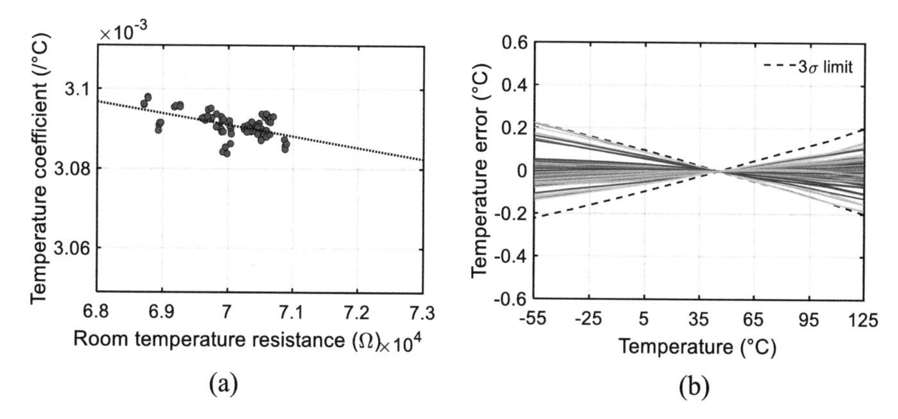

Fig. 3.45 (**a**) Correlation between R_0 and *TC* (**b**) sensor inaccuracy after a correlation-assisted 1-point trim and systematic nonlinearity correction, with 4 outliers excluded

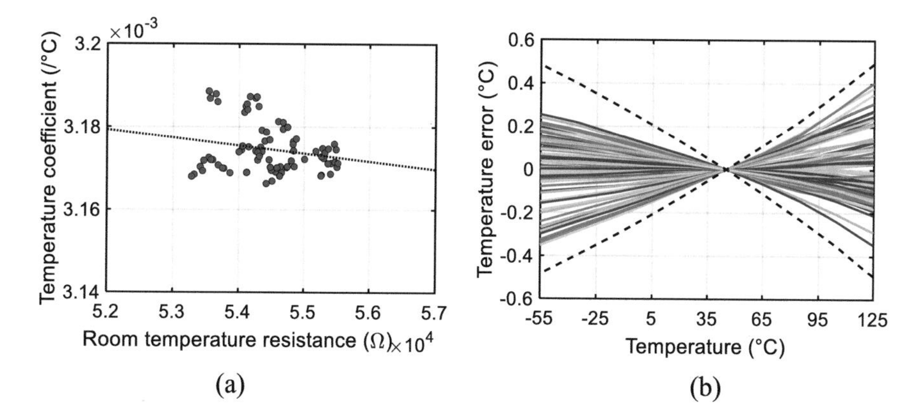

Fig. 3.46 (**a**) Correlation between R_0 and *TC* (**b**) sensor inaccuracy after a correlation-assisted 1-point trim and systematic nonlinearity correction for sensors in another batch

To investigate which model is better, an almost identical WB sensor design has been fabricated in another batch and then characterized. Figure 3.46 shows its R_0-TC correlation and the temperature inaccuracy. Compared to Fig. 3.45, the TC spread is larger, and the 1-point-trimmed inaccuracy is degraded to about 0.5 °C (3σ). Even assuming no R_0-TC correlation, the 1-point-trimmed inaccuracy is only 5% worse. It can be concluded that, the R_0-TC correlation is either nonexistent or too weak to be relied on for the narrow (0.5 μm) silicided poly resistor.

3.5.2.3 Comparison to Implementation II

Compared to the second prototype, the third implementation achieves 3× better relative inaccuracy with the same chip area. This is mainly due to the improved opamp design and use of system-level chopping.

3.6 Comparisons and Concluding Remarks

In this chapter, three Wien bridge–based temperature sensor prototypes were presented. All the sensors use a second-order phase-domain ΔΣ ADC to digitize the temperature-dependent phase shift of the WB sensor. The characteristics of these sensors are summarized and compared to the state of the art (in 2020) in Table 3.2.

Compared to the first prototype, the readout circuits in the next two implementations were further optimized, resulting in less chip area and, finally, in better relative inaccuracy. In the most recent implementation, the 0.12 mm^2 WB sensor achieves a relative inaccuracy of 0.03% with only two trimming points, and a 1/*f* noise corner of below 10 mHz. However, the resolution FoM of such sensors (>100 fJ·K^2) is far larger than the theoretical value (2.3 fJ·K^2), which is mainly due to the power and noise from the readout circuit.

Similar or even better energy efficiency (<100 fJ·K^2) can be achieved with a frequency-based readout [13–15], that is, a frequency-locked loop (FLL) followed by a frequency-to-digital converter (FDC). Like a phase-domain ΔΣ ADC, the power consumption and noise performance of an FLL are dominated by that of its integrator (Fig. 2.9). By using filtering [13] or sampling [14] techniques, however, the integrator's AC input can be greatly suppressed, allowing its design to be optimized for noise. This is not the case for the sensors presented in this chapter, since their integrators must be designed to handle the maximum output current of their

Table 3.2 Performance summary of the WB sensors presented in this Chapter and that of other state-of-the art RC-based temperature sensors

	JSSC'20 [13]	ISSCC'20 [14]	Implementation I JSSC'18	Implementation II ISSCC'19	Implementation III ISSCC'21
Sensor type	SC-WhB	RC	WB	WB	WB
ADC type	FLL+FDC	FLL+FDC	ΔΣM	ΔΣM	ΔΣM
Technology	180 nm	65 nm	180 nm	180 nm	180 nm
Area [mm^2]	0.72	0.0088	0.72	0.12	0.12
Temp. Range [°C]	−40—85	−30—90	−40—85	−40—180	−55—125
3σ Inaccuracy [°C] (trimming points)	0.55 (2)	0.32[a] (2)	0.03 (2[b])	0.1 (2[b])	0.03 (2[b])
Relative inaccuracy	0.88%	0.52%	0.05%	0.1%	0.03%
Power [μW]	15.6	45	160	52	66
Conv. time [ms]	1	1	5	10	10
Resolution [mK]	2	1.43	0.44	0.46	0.45
Res. FoM [fJ·K^2]	62	92	150	110	130

[a]Min/Max. [b]1st-order fit

WB sensors. Another way to suppress the sensor's power is to digitize the front-end output before entering the FLL [15]. However, due to the large comparator noise, this typically results in a slightly worse energy efficiency.

Despite a better (or similar) FoM, the accuracy of FLL-based sensors is currently significantly worse, possibly due to the charge injection of the switched-capacitor circuits used in their RC front-ends. Furthermore, such sensors also require an FDC to digitize the frequency output, whose extra power consumption and area is not always presented.

References

1. M. Shahmohammadi, K. Souri, K.A.A. Makinwa, A resistor-based temperature sensor for MEMS frequency references, in *Proc. ESSCIRC*, (2013), pp. 225–228
2. C.P.L. van Vroonhoven, K.A.A. Makinwa, A CMOS temperature-to-digital converter with an inaccuracy of ±0.5°C (3σ) from −55 to 125°C, in *IEEE ISSCC Dig. Tech. Papers*, (2008), pp. 576–577
3. K.A.A. Makinwa, Smart temperature sensors in standard CMOS, in *(Proc. Eurosensors) Procedia Engineering*, (2010), pp. 930–939
4. K.Y. Nam, S.-M. Lee, D.K. Su, B.A. Wooley, A low-voltage low-power sigma-delta modulator for broadband analog-to-digital conversion. IEEE J. Solid State Circuits **40**(9), 1855–1864 (2005)
5. C.P.L. van Vroonhoven, D. d'Aquino, K.A.A. Makinwa, A thermal-diffusivity-based temperature sensor with an untrimmed inaccuracy of ±0.2°C (3σ) from −55°C to 125°C, in *ISSCC Dig. Tech. Papers*, (2010), pp. 314–315
6. A. Hastings, *The Art of Analog Layout* (Prentice Hall, Englewood Cliffs, NJ, 2001)
7. A. Heidari, G. Wang, K. Makinwa, G.C.M. Meijer, A BJT-based CMOS temperature sensor with a 3.6pJK^2 resolution FOM, in *IEEE ISSCC Dig. Tech. Papers*, (2014), pp. 224–225
8. C. Weng, C. Wu, T. Lin, A CMOS thermistor-embedded continuous-time Delta-Sigma temperature sensor with a resolution FoM of 0.65 pJ °C^2. IEEE J. Solid State Circuits **50**(11), 2491–2500 (2015)
9. P. Park, D. Ruffieux, K.A.A. Makinwa, A thermistor-based temperature sensor for a real-time clock with ±2 ppm frequency stability. IEEE J. Solid State Circuits **50**(7), 1571–1580 (2015)
10. U. Ausserlechner et al., Drift of magnetic sensitivity of smart Hall sensors due to moisture absorbed by the IC-package, in *Proc. IEEE Sensors*, (2004), pp. 455–458
11. B. Yousefzadeh et al., A BJT-based temperature sensor with a packaging-robust inaccuracy of ±0.3°C (3σ) from −55°C to +125°C After Heater-Assisted Voltage Calibration, in *IEEE ISSCC Dig. Tech. Papers*, (2017), pp. 162–163
12. R. Wu, K.A.A. Makinwa, J.H. Huijsing, A chopper current-feedback instrumentation amplifier with a 1 mHz 1/*f* noise corner and an AC-coupled ripple reduction loop. IEEE J. Solid State Circuits **44**(12), 3232–3243 (2009)
13. H. Jiang, C.-C. Huang, M.R. Chan, D.A. Hall, A 2-in-1 temperature and humidity sensor with a single FLL Wheatstone-bridge front-end. IEEE J. Solid State Circuits **55**(8), 2174–2185 (2020)
14. A. Khashaba et al., A 0.0088mm^2 resistor-based temperature sensor achieving 92fJ·K^2 FoM in 65nm CMOS, in *IEEE Dig. Tech. Papers*, (2020), pp. 60–61
15. Y. Lee et al., A 5800-μm^2 resistor-based temperature sensor with a one-point trimmed inaccuracy of ±1.2°C (3σ) from −50°C to 105°C in 65-nm CMOS. Solid-State Circuits L. **2**(9), 67–70 (2019)
16. S. Pan, Y. Luo, S.H. Shalmany, K.A.A. Makinwa, A resistor-based temperature sensor with a 0.13 pJ·K2 resolution FoM. IEEE J. Solid-State Circuits **53**(1), 164–173 (2018)

17. S. Pan, Ç. Gürleyük, M.F. Pimenta, K.A.A. Makinwa, A 0.12mm[2] Wien-bridge temperature sensor with 0.1 C (3σ) inaccuracy from −40 C to 180 C, in *ISSCC Dig. Tech. Papers*, (2019), pp. 184–186
18. S. Pan, J.A. Angevare, K.A.A. Makinwa, A hybrid thermal-diffusivity/resistor-based temperature sensor with a self-calibrated inaccuracy of 0.25°C (3σ) from −55 °C to 125 °C, in *ISSCC Dig. Tech. Papers*, (2021), pp. 78–79

Chapter 4
Wheatstone Bridge–Based Temperature Sensors

4.1 Introduction

In contrast to Wien bridge sensors, on-chip Wheatstone bridge (WhB) sensors are less accurate, due to the lack of a stable reference resistor. However, because of the extra sensitivity that can be achieved by using resistors with opposite temperature coefficients (TCs), they can achieve better energy efficiency.

As presented in Chap. 2, the TC of silicided resistors is positive (TC_p), while that of some nonsilicided poly resistors is negative (TC_n). In the chosen 0.18 μm process, $TC_p = 0.285\%/°C$ for the silicided-p-poly resistor, while $TC_n = -0.152\%/°C$ for the nonsilicided n-poly resistor. According to Eq. (2.1), the theoretical FoM of the resulting Wheatstone bridge will then be ~1.7 fJK^2. Assuming the readout circuit dissipates the same power and contributes the same noise as the bridge, the practical FoM limit will then be ~7 fJ·K^2.

Like Chap. 3, this chapter starts with a discussion of some general design choices, including the choice of readout circuit topology and the trimming method. Afterward, four Wheatstone bridge sensor implementations are presented based on their publication sequence. By systematically optimizing their circuit design, their resolution FoM improves from 65 fJ·K^2 (the first implementation) to 10 fJ·K^2 (the fourth implementation).

S. Pan, K. A. A. Makinwa, *Resistor-based Temperature Sensors in CMOS Technology*, ACSP · Analog Circuits and Signal Processing,
https://doi.org/10.1007/978-3-030-95284-6_4

4.2 General Design Choices

4.2.1 Traditional Readout Versus Direct Readout

Traditionally, a Wheatstone bridge sensor output is read out by digitizing its open-circuit voltage, as shown in Fig. 4.1a, where R_p and R_n are resistors with positive and negative TCs, respectively. An instrumentation amplifier (IA) can be used to match the output voltage and impedance of the bridge to the input range and impedance of the ADC, resulting in good energy efficiency. Compared to the classic 3-opamp IA [1], a significant improvement in energy efficiency can be achieved by using a Current Feedback Instrumentation Amplifier (CFIA) [2] or a Capacitively Coupled Instrumentation Amplifier (CCIA) [3].

One drawback of this traditional topology is that the sensor's accuracy will be limited by the combined gain errors of the IA and the ADC. Moreover, the sensor's output range is limited to V_{DD} over the gain of the IA. To effectively suppress the input-referred noise of the ADC, the IA's gain is usually larger than 10 [4], which sets a maximum WhB output of 0.1 V_{DD}. In the chosen process, however, the WhB output range is roughly 14% over the industrial temperature range (−40 °C to 85 °C) or almost 20% over the military range (−55 °C to 125 °C). Consequently, the energy efficiency of this traditional topology cannot be optimized.

A simpler and more accurate way of reading out a WhB is by directly balancing it with a resistive DAC, and hence, nulling its output current. As shown in Fig. 4.1b, the required feedback loop can be conveniently realized as a continuous-time delta-sigma modulator (CTΔΣM) [5], where the modulator's bitstream output D_{out} drives the resistive DAC to null the difference I_{err}(T) between the output currents of the DAC and the bridge. The resulting bitstream average will then be proportional to the amount of parallel resistance required to balance the bridge. If R_{DAC} is also an R_n-type resistor, the bitstream average will be solely determined by the temperature-dependent values of R_n and R_p resistors. The system is also robust to supply voltage

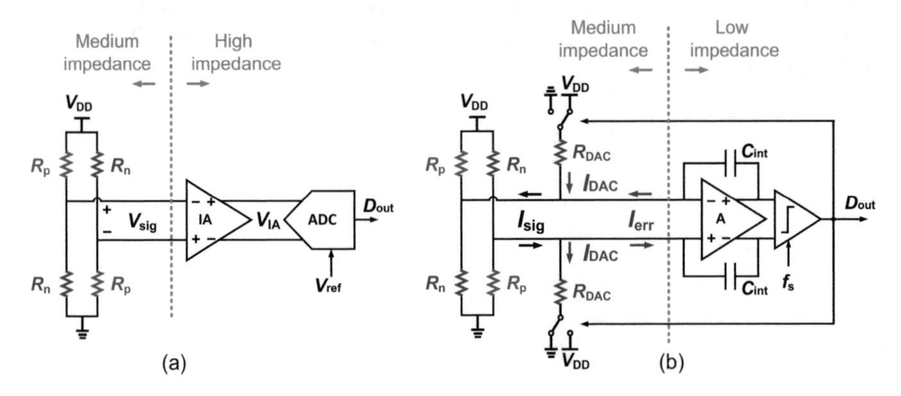

Fig. 4.1 (**a**) Conventional WhB readout using IA and ADC and (**b**) CTΔΣM WhB readout with a resistive DAC

variations, since the DAC and the bridge share the same supply. Moreover, the WhB output range has no limitation, as the integration capacitor (C_{int}) can be scaled to limit the maximum output voltage of the integrator. In this chapter, this direct readout scheme is used in all four implementations.

4.2.2 Nonlinearity and Trimming

As in Chap. 2, the temperature dependence of the various resistors in the WhB and the DAC can be modeled as:

$$\begin{aligned} R_p(T) &= R_p(T_0)\cdot\left(1+TC_{p1}\cdot\Delta T+TC_{p2}\cdot\Delta T^2\right) \\ R_n(T) &= R_n(T_0)\cdot\left(1+TC_{n1}\cdot\Delta T+TC_{n2}\cdot\Delta T^2\right) \\ R_{DAC}(T) &= R_{DAC}(T_0)\cdot\left(1+TC_{n1}\cdot\Delta T+TC_{n2}\cdot\Delta T^2\right). \end{aligned} \quad (4.1)$$

Here R_p (T_0), R_n (T_0), and R_{DAC} (T_0) are the resistances at a reference temperature T_0, while TC_{p1}, TC_{n1}, are their first-order TCs, TC_{p2} and TC_{n2} are their second-order TCs, and ΔT is the temperature with respect to T_0. Noting that the active integrator virtually shorts the bridge's output terminals to $V_{DD}/2$, while the modulator ensures that the integrator's average input current I_{err} is zero, the bitstream average μ_{ADC} can be expressed as:

$$\begin{aligned} \mu_{ADC} &= \frac{I_{sig}(T)}{I_{DAC}(T)} = \frac{1/R_p(T)-1/R_n(T)}{1/R_{DAC}(T)} \\ &= \frac{R_{DAC}(T_0)}{R_p(T_0)}\cdot\frac{\left(1+TC_{n1}\cdot\Delta T+TC_{n2}\cdot\Delta T^2\right)}{\left(1+TC_{p1}\cdot\Delta T+TC_{p2}\cdot\Delta T^2\right)}-\frac{R_{DAC}(T_0)}{R_n(T_0)} \\ &= \frac{R_{DAC}(T_0)}{R_p(T_0)}\cdot f_{pn}(\Delta T)-\frac{R_{DAC}(T_0)}{R_n(T_0)}. \end{aligned} \quad (4.2)$$

Since the TC spread of CMOS resistors is much smaller than their nominal resistance spread (Chap. 3, Sect. 3.3.2.2), the function f_{pn}, can be approximated as a constant, but nonlinear, function of temperature. The ratio R_{DAC}/R_p involves different types of resistors and so will spread significantly, while the ratio R_{DAC}/R_n involves the same type of resistors and so should spread less. So a one-point trim is definitely required to compensate for the spread of R_{DAC}/R_p. A two-point trim, on the other hand, will compensate for the spread of R_{DAC}/R_n, as well as some spread due to the varying TCs.

According to (4.2), the mapping of the main spread component R_{DAC}/R_p to μ_{ADC} is linear, so unlike WB sensors (Chap. 3, Sect. 3.2.3.2), no nonlinearity correction polynomial is required before performing a two-point trim. This greatly simplifies the process of calibrating a WhB sensor.

4.3 Implementation I, Proof of Concept[1]

In this first design, a WhB sensor is built by reusing circuit blocks from the first WB sensor design (Chap. 3, Sect. 3.3). It achieves a resolution FoM of 65 fJ·K^2, which is better than that of all the WB sensors. However, due to the lack of optimization, this FoM is still almost 10× worse than the practical FoM limit.

4.3.1 Circuit Implementation

As shown in Fig. 4.2, the proposed Wheatstone bridge temperature sensor consists of silicided p-poly (s-p-poly, R_p = 105 kΩ, $TC{\approx}0.285\%/°C$) and a nonsilicided n-poly (R_n = 95 kΩ, $TC{\approx} - 0.152\%/°C$) resistors. No investigate the performance of the nonsilicided p-poly resistor, a bridge made from s-p-poly and p-poly resistors ($TC{\approx} - 0.02\%/°C$) was also realized. To reuse the same readout circuit, its bridge resistances (R_p = 67.5 kΩ, R_n = 64 kΩ) are chosen to ensure the same maximum error current levels over the industrial temperature range from −40 °C to 85 °C.

As in Sect. 4.2.1, the Wheatstone bridge sensors are read out by connecting them to the virtual ground of the first integrator of a CTΔΣ-ADC (Fig. 4.2). The modulator's resistive DAC (R_{DAC} = 140 kΩ, made from the same material as R_n) will then null their output current. The DAC resistors are switched to either supply rails or left floating, so that this differential sensor has 4 DAC resistors. As will be shown in the second design, however, the same function can be performed by switching two DAC resistors between the supply rails.

In this work, in contrast to previous work [5] shown in Fig. 4.3, the use of a feed-forward architecture helps to suppress the swing of the first integrator. More

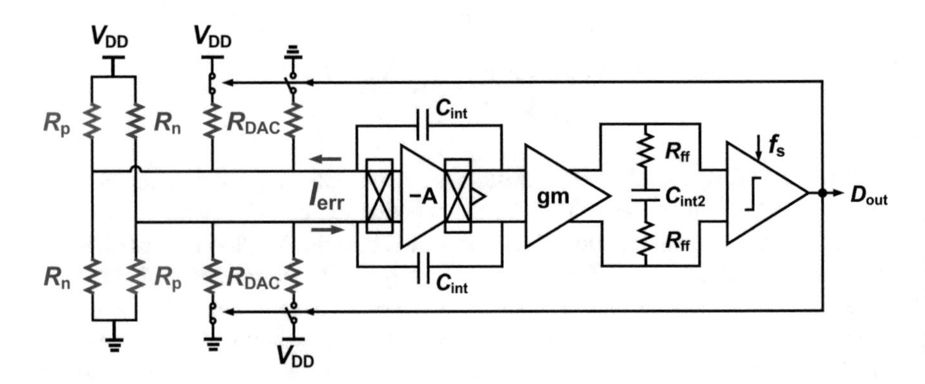

Fig. 4.2 System block diagram of the first WhB sensor implementation

[1] Pan et al. [7]

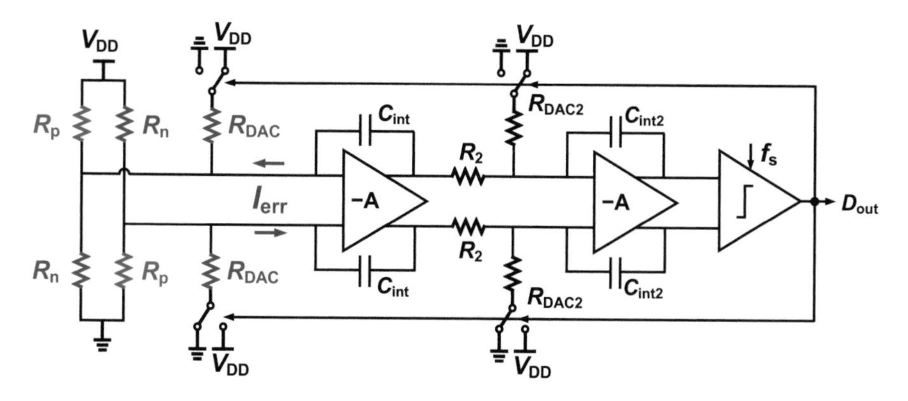

Fig. 4.3 System block diagram of a previous Wheatstone bridge sensor [5]

importantly, the offset and 1/*f* noise of the first integrator are suppressed by chopping. To avoid aliasing high-frequency quantization noise at the chopping transitions [6], the chopping frequency is the same as the sampling frequency (500 kHz).

Both the amplifiers used in the first and second integrators were reused from the first WB prototype (Chap. 3, Sect. 3.3). From simulations, the first and second integrators dissipate 100 μW and 7 μW, respectively, from a 1.8 V supply. The bridges dissipate 32 μW (s-p/n-poly) and 25 μW (s-p/p-poly).

4.3.2 Measurement Results

The two Wheatstone bridge sensors were fabricated side by side, as shown in Fig. 4.4. They share the same clock and constant-g_m biasing circuits and each occupies 0.72 mm^2, which is dominated by the large capacitors (2 × 180 pF) of the first integrator. For flexibility, the sinc^2 decimation filter is realized off-chip.

4.3.2.1 Calibration and Inaccuracy

Twenty samples from one wafer were characterized in ceramic packages from −40 °C to 85 °C in a temperature-controlled oven. The sensor characters are shown in Fig. 4.5. After a first-order fit, the measured nonlinearities are quite systematic (Fig. 4.6), and, like those shown in Chap. 3, can be removed by a fixed polynomial. The resulting spread is below 0.07 °C (3σ) for the s-p-poly/p-poly sensor, and below 0.10 °C (3σ) for the s-p-poly/n-poly sensor (Fig. 4.7). At room temperature, the power supply sensitivity of both sensors is less than 20 mK/V.

Fig. 4.4 System block diagram of the first WhB sensor implementation

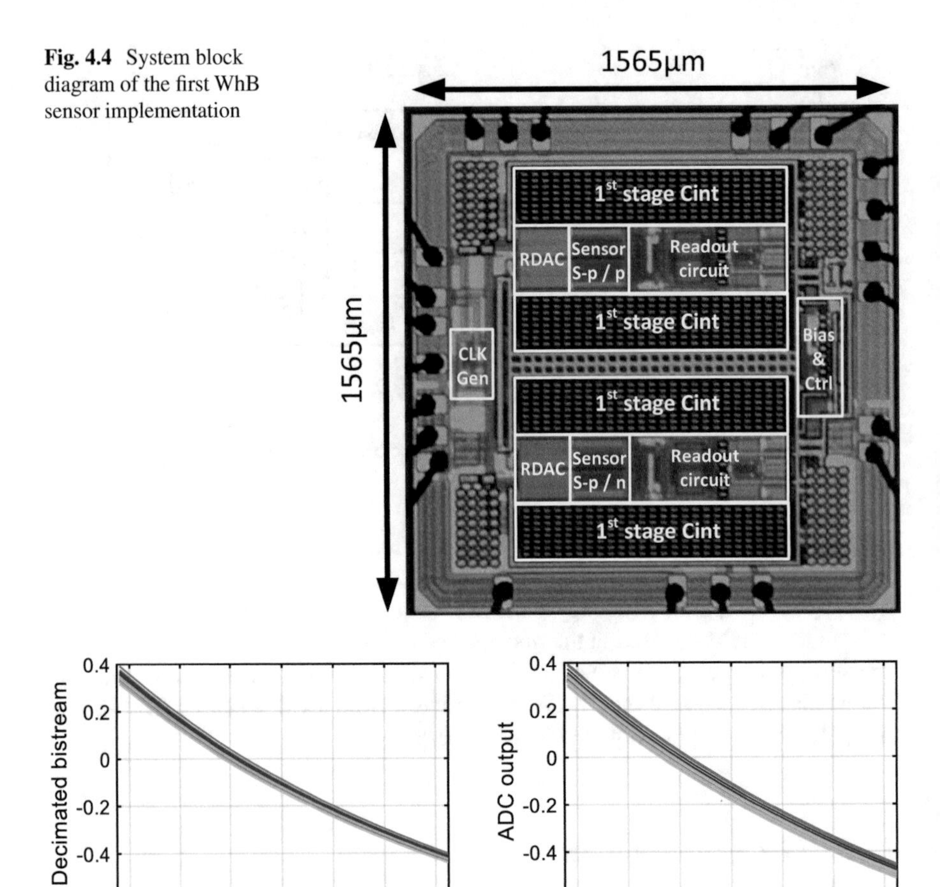

Fig. 4.5 Output versus temperature of (**a**) s-p-poly/n-poly WhB sensors and (**b**) s-p-poly/p-poly WhB sensors

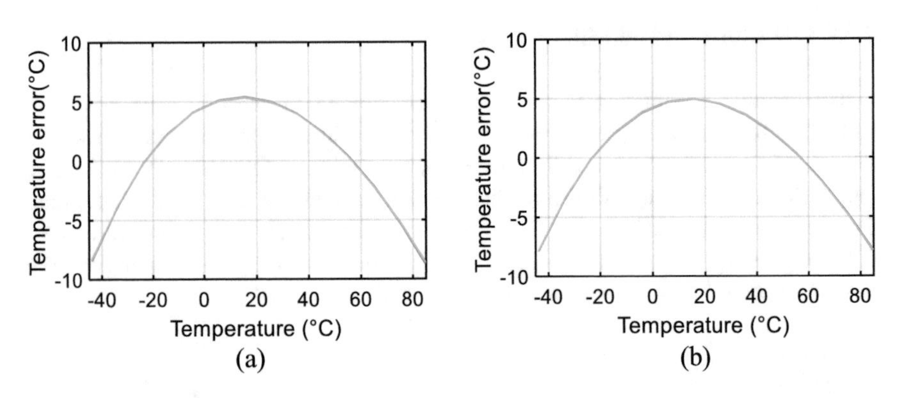

Fig. 4.6 Temperature error of (**a**) s-p-poly/n-poly WhB sensors and (**b**) s-p-poly/p-poly WhB sensors after individual first-order fit

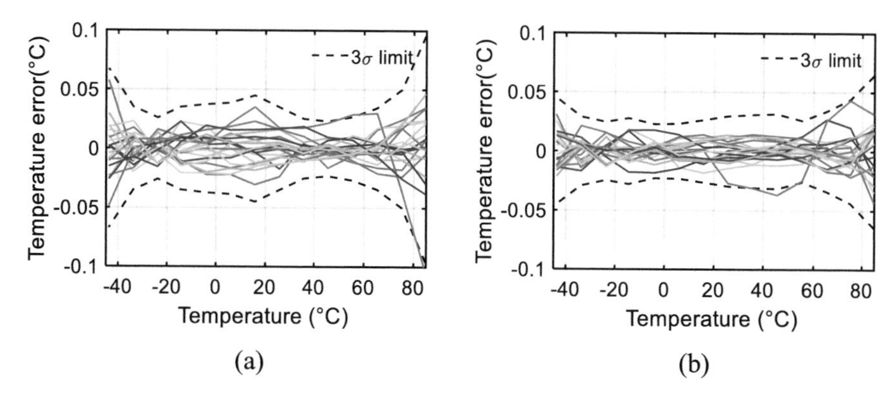

Fig. 4.7 Residual temperature error of (**a**) s-poly WhB sensors and (**b**) s-diffusion WhB sensors after individual first-order fit and systematic nonlinearity removal

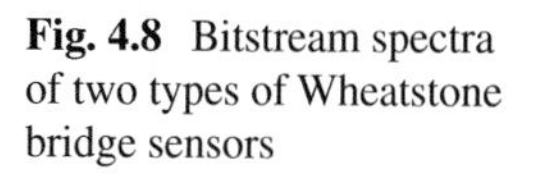

Fig. 4.8 Bitstream spectra of two types of Wheatstone bridge sensors

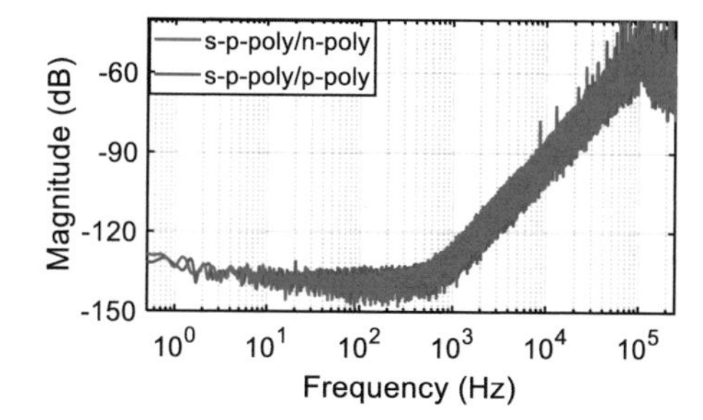

4.3.2.2 Resolution and FoM

The bitstream spectra of the sensors are shown in Fig. 4.8. The observed $1/f$ noise is mainly due to the nonsilicided poly resistors, resulting in a 10 Hz corner frequency for both sensors. In [7], it was reported that the s-p-poly/n-poly sensor achieves a thermal-noise limited resolution of 164 μK in a conversion time of 10 ms (Fig. 4.9), which corresponds to a resolution FoM of 49 fJ·K^2. However, just like that in [8], the use of a two-sample Allan deviation underestimates the sensor's $1/f$ noise. With a more realistic estimation using standard deviation in a 1 s interval (Chap. 3, Sect. 3.3.2.1), the resolution becomes ~15% worse, and the corresponding FoM becomes 65 fJ·K^2. Although the s-p-poly/p-poly sensor achieves a similar level of resolution with the same conversion time, its resolution FoM is ~10% worse due to its larger power consumption.

4.3.2.3 Comparison with Prior Art

Compared to [5], the energy efficiency of the proposed s-p-poly/n-poly Wheatstone bridge sensor is improved by 10×, which is mainly due to the increased sensitivity from silicided-poly resistors and the reduced $1/f$ noise via chopping. As for the

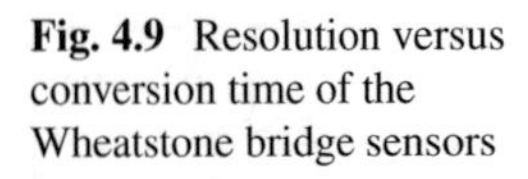

Fig. 4.9 Resolution versus conversion time of the Wheatstone bridge sensors

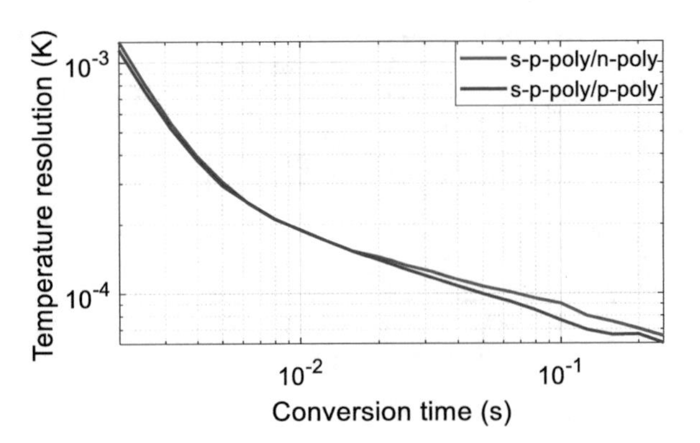

s-p-poly/p-poly sensor, its resolution FoM is ~10% worse than the s-p-poly/n-poly sensor, but the inaccuracy is somewhat better.

The sensor requires further optimization: its area is ~70% larger compared to [5], and the efficiency gap between the achieved FoM (65 fJ·K^2) and the practical FoM limit (7 fJ·K^2) is almost 10×.

4.4 Implementation II, Smaller Area and Better FoM[2]

In this second implementation, a multi-bit CTΔΣ-ADC is proposed to replace the single-bit ADC used in the first Wheatstone bridge sensor prototype. It achieves a resolution FoM of 40 fJ·K^2 and an active chip area of 0.25 mm^2.

4.4.1 System-Level Design

Just like the first Wien bridge implementation, the first Wheatstone bridge prototype consumes most of its power/area in the first-stage amplifier/integration capacitor. These are all limited by the large variation in the WhB output signal I_{sig} over PVT. As shown in Fig. 4.10a, the first stage of the CTΔΣ-ADC is essentially an active-RC integrator, where the WhB is modeled by a source resistor driven by a temperature-dependent voltage $V_{in}(T)$. This has to be compensated by the output current I_{DAC} of a 1-bit DAC, resulting in an even larger error current I_{err} flowing into the first integrator, as shown in Fig. 4.10b.

A multi-bit resistor DAC (N > 1) can be used to reduce the magnitude of I_{err} (Fig. 4.10c). Since most of I_{sig} will then be compensated by I_{DAC}, the first integrator's supply current, as well as the size of its integration capacitors, can be significantly reduced.

[2] Pan and Makinwa [22].

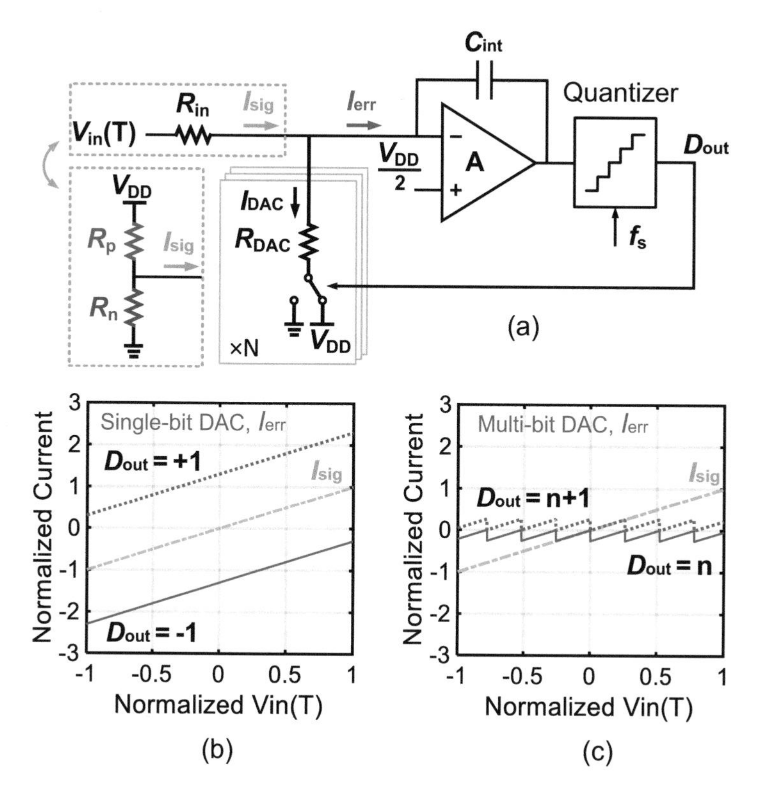

Fig. 4.10 (**a**) A CTΔΣ-ADC based on an active integrator. (**b**) Error current of a single-bit CTΔΣ-ADC. (**c**) Error current of a multi-bit CTΔΣ-ADC

Nonlinearity is a key challenge in multi-bit ΔΣMs. For CTΔΣMs with resistive DACs, the two major contributors are RDAC mismatch and the nonlinearity of the first integrator. RDAC mismatch can be sufficiently suppressed by careful layout and dynamic element matching (DEM). The nonlinearity of the first integrator, however, is more problematic. The input impedance of the CTΔΣM becomes signal dependent, and so will the DAC value. As in [9], this creates in-band noise (IBN) that cannot be mitigated by DEM. Increasing the linearity of the first-stage amplifier would help, but this usually comes at the expense of higher power dissipation.

This problem can be solved using a zoom-ADC [10]. During the fine conversion of a zoom ΔΣM, however, only two levels of its multi-bit DAC will be used. As a result, the DAC will still appear to be perfectly linear even in the presence of integrator nonlinearity, and so no quantization noise folding will occur [11].

The proposed zoom CTΔΣM digitizes the temperature-dependent ratio $X = I_{sig} / (2I_{DAC})$ in two steps, as illustrated in Fig. 4.11a, for the case of a first-order modulator. First, a coarse SAR conversion determines the integer part n of X. Then, the fractional part μ is determined by a fine ΔΣ conversion. Compared to conventional multi-bit ΔΣMs, a zoom-ADC obviates the complexity of a multi-bit quantizer and

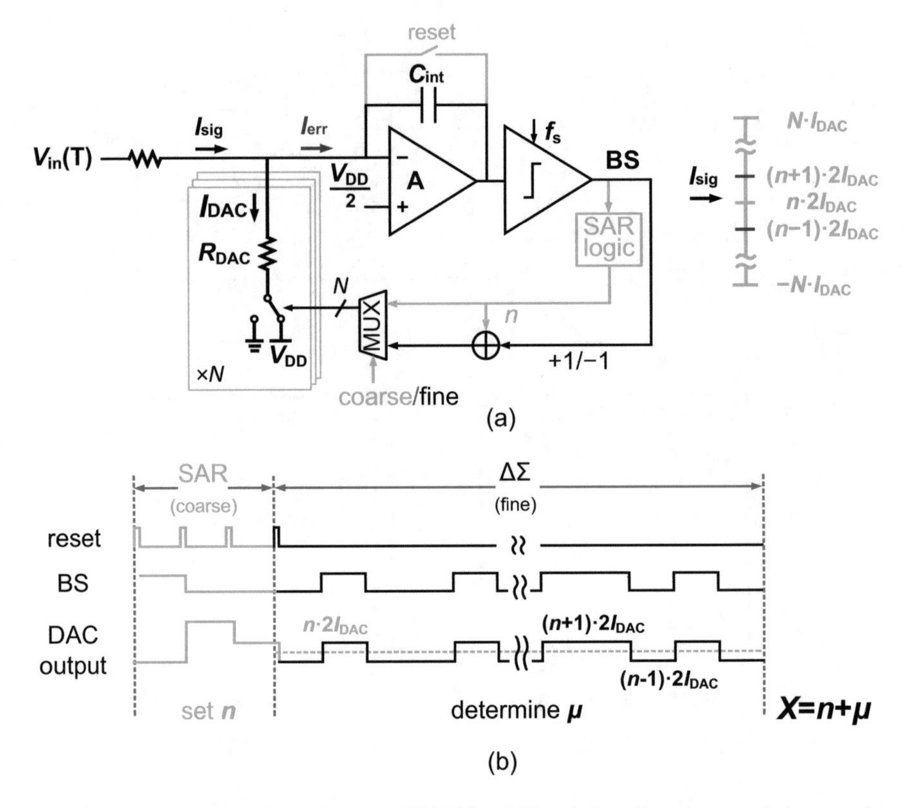

Fig. 4.11 (**a**) Schematic of a static zoom CTΔΣM and (**b**) a timing diagram example in case of a 3-bit DAC

only requires a single-bit comparator, a reset switch for the first integrator, and some logic.

During the coarse conversion, the first integrator is used as a pre-amplifier for the comparator [12]. Each step of the SAR conversion then consists of choosing a DAC code, resetting the first integrator, and then integrating the resulting error current for one clock cycle. The polarity of the result is detected by the comparator and used to determine the next DAC code to test. To absorb potential errors from the SAR conversion and ensure that μ lies in the modulator's stable input range, over-ranging is used. This is implemented by switching the DAC between the codes $n - 1$ and $n + 1$ during the delta-sigma phase (Fig. 4.11b). This significantly reduces the current I_{err} flowing into the integrator so the linearity and power dissipation of the first integrator can be significantly relaxed.

4.4.2 Circuit Implementation

4.4.2.1 Wheatstone Bridge and DAC

In contrast to the first Wheatstone bridge prototype, this sensor is designed to operate over the military temperature range (−55 °C to 125 °C). As a result, the bridge is unbalanced at room temperature (RT, about 25 °C), with R_p = 100 kΩ and R_n = 80 kΩ. With a 1.8 V supply, this results in $|I_{sig}| < 7$ μA over PVT, which requires a minimum DAC resistance of 120 kΩ. The minimum width imposes a trade-off between the number of DAC bits and the minimum possible DAC area. To ensure that the areas of the DAC and the integrating caps C_{int} are roughly equal, a 3-bit DAC was chosen, with unit elements of 960 kΩ. As in [12], an extra half-LSB unit element is used at the end of the coarse conversion to determine the optimal choice of the references used in the fine conversion.

4.4.2.2 Zoom ADC

Figure 4.12 shows the circuit diagram of the zoom CTΔΣ-ADC. To achieve high resolution in a reasonable conversion time, a second-order modulator was chosen with a feedforward architecture to reduce the swing at the output of the first integrator and thus further reduce the size of C_{int} (~48 pF). As in the first Wien bridge sensor prototype, the loop is stabilized by a zero realized by R_{ff} at the output of the second integrator.

Since the first integrator's nonlinearity will not increase IBN, it was optimized mainly for noise. It employs an energy-efficient current-reuse OTA [13], rather than the two-stage opamp used in the first Wheatstone bridge prototype. High-V_T input

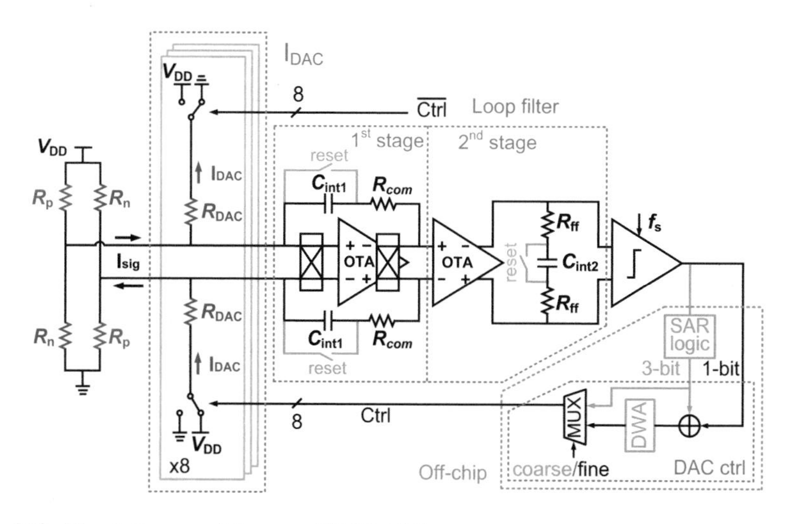

Fig. 4.12 Circuit diagram of the zoom CTΔΣ-ADC

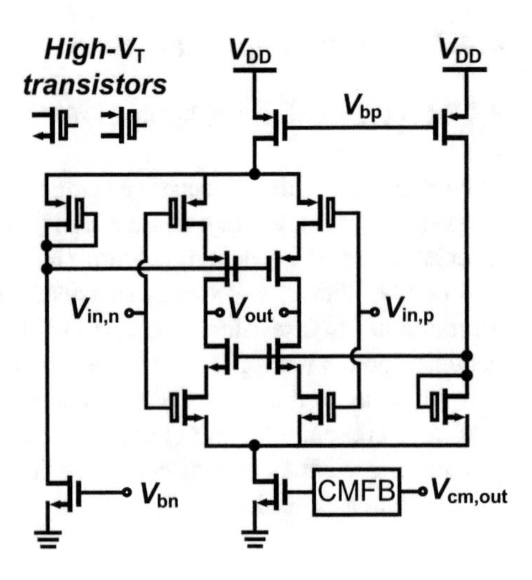

Fig. 4.13 Schematic of the first-stage OTA

transistors are used to achieve a reasonable output swing (~0.9 V at RT), as shown in Fig. 4.13. To improve modulator stability, the zero of the OTA-based integrator is compensated by inserter $R_{com} \approx 1/g_m$ in series with C_{int}. To suppress its offset and $1/f$ noise while avoiding quantization noise fold-back, the OTA is chopped at the CTΔΣM's sampling frequency ($f_s = 500$ kHz). It achieves over 80 dB of DC gain, a GBW product of ~20 MHz with a 1 pF load, and consumes 22 μW at RT, which is about 60% of the power dissipated by the bridge. The second stage is based on a source-degenerated cascaded telescopic OTA. It has a DC gain of 80 dB and dissipates 3 μW at RT.

For flexibility, the SAR and DWA logic are implemented off-chip. Since the SAR conversion only involves 3-bits, its duration and power overhead are negligible, and the energy efficiency of the bridge readout is basically defined by the fine conversion. Simulations show that, if implemented on-chip, the SAR and DWA logic would consume less than 1 μW and less than 0.01 mm² area, which are negligible compared to the other circuit blocks.

4.4.2.3 Nonlinearity and Segment Averaging

Although the nonlinearity of the first integrator does not impact the IBN of a zoom ADC, it does impact its INL. The main source of nonlinearity is the signal-dependent g_m of the current-reuse OTA, which can be modeled by the addition of a third-order term g_{m3}.

The OTA's nonlinearity will cause errors in the bitstream average μ obtained after the fine conversion. These will be a weighted average of the associated errors in the two possible values of the first integrator's input current I_{err}. When $\mu = 0$, however, the bitstream output BS will toggle between +1 and − 1 with equal

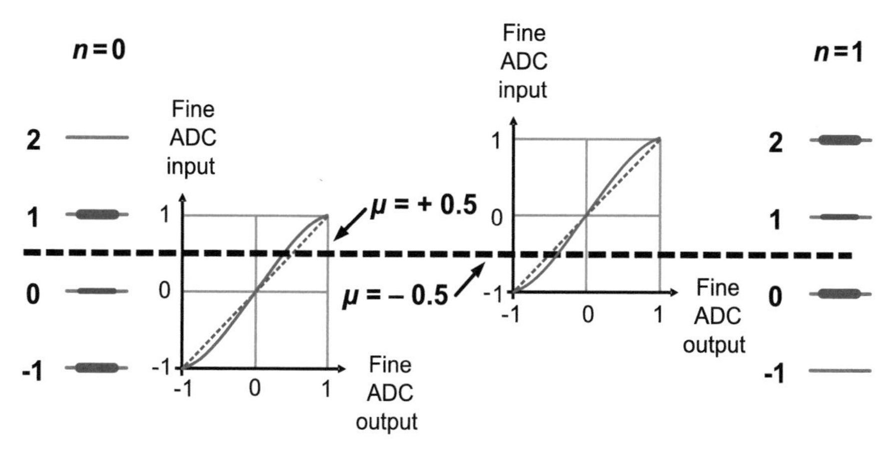

Fig. 4.14 Readout error of the opposite sign with two possible coarse codes

probability, and since I_{err} $(BS = +1) = -$ I_{err} $(BS = -1)$, the resulting error in μ will be zero. This will also be the case at the extremes of the modulator's input range, because the bridge's output current I_{sig} will then be exactly canceled by I_{DAC}, and so $I_{err} = 0$. Apart from these three cases, the error of the fine ADC will be non-zero. As shown in Fig. 4.14, the result is a sinusoidal error curve centered on $\mu = 0$.

Choosing the range of the fine conversion to be exactly equal to two steps of the coarse conversion (2-LSB over-ranging) means that there are two different ways to convert a given input current, each corresponding to a different coarse code n. Ideally, the zoom ADC's output X would be the same in both cases. In the presence of OTA nonlinearity, however, there will be an error in μ, which will be of opposite polarity in the two cases. As shown in Fig. 4.15, this means that at the coarse code transitions, that is, when $\mu = \pm 0.5$, the error in X will abruptly change polarity. Simulations show that the jumps in X at RT can be as large as 0.1 °C, which is significantly larger than the sensor's expected resolution.

Noting that the errors associated with the two possible n/μ combinations are of opposite polarity, they can be mitigated by simply averaging the values of X obtained from two such conversions, as shown in Fig. 4.16a. Simulations show that this approach can reduce the error by about 8×, to about ±5 mK. This approach translates to considerable power savings. Without this segment averaging technique, the bias current of the first integrator's OTA would have to be increased by about 2× to obtain similar linearity.

Although the stable input range of a second-order ΔΣM corresponds to $-1 < \mu < 1$ for DC input signals [14], its quantization noise becomes quite large when $|\mu| \sim 1$. To avoid degrading the sensor's resolution in such cases, segment averaging is disabled when $1 - |\mu| < 0.05$. As shown in Fig. 4.16b, this will have little effect on the sensor's linearity, since the nonlinearity is anyway quite small in these cases and the transitions are blurred by the presence of thermal noise.

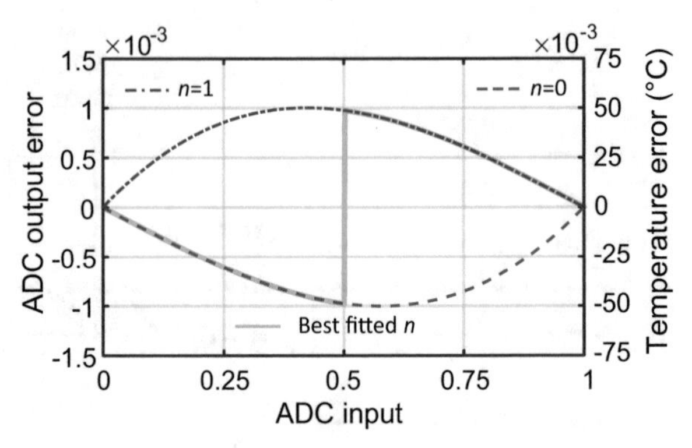

Fig. 4.15 Temperature error jump when changing n

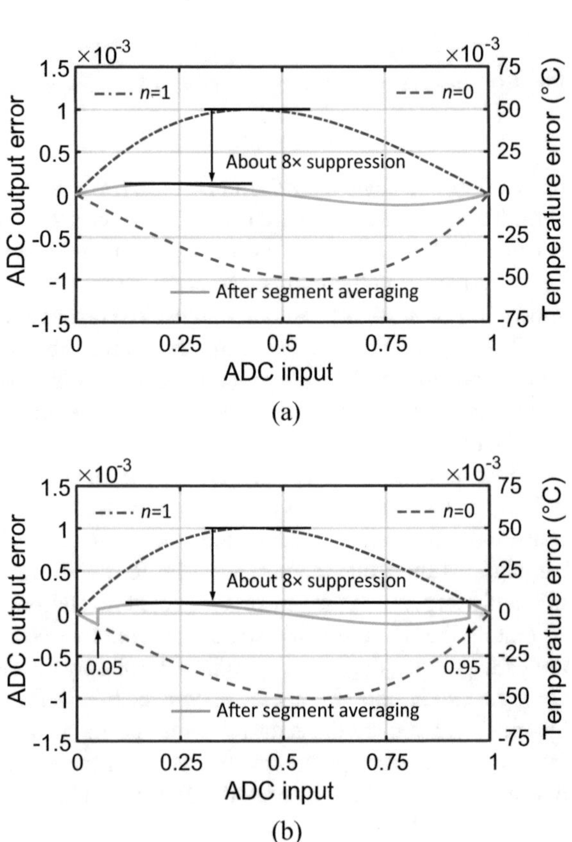

Fig. 4.16 Nonlinearity suppression using segment averaging (**a**) without a threshold (**b**) with a threshold of 0.05

4.4.3 Measurement Results

The sensor is realized in the same standard 0.18 μm CMOS process, with a dimension of 615 μm × 410 μm (Fig. 4.17). At RT, it draws 52 μA from a 1.8 V supply, with over half of this dissipated in the WhB and the DAC. About 15% of the active

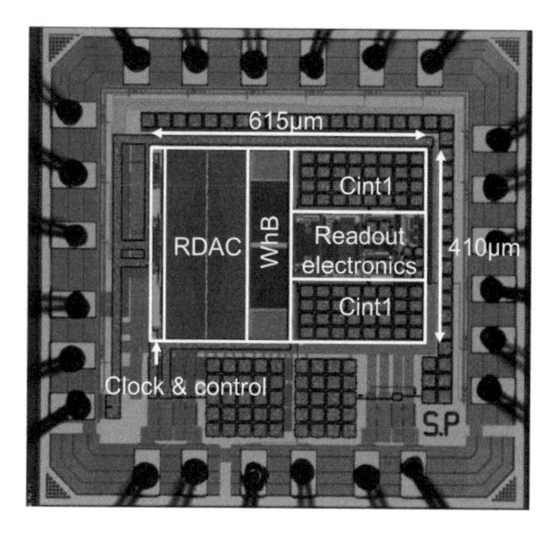

Fig. 4.17 Die micrograph of the second Wheatstone sensor implementation

area is occupied by the WhB, 30% by the DAC resistors, and another 30% by the integration capacitors of the first stage. For supply voltages varying from 1.6 V to 2.0 V, the sensor's supply sensitivity is 0.02 °C/V. An off-chip sinc^2 filter is used to decimate the sensor's bitstream output.

4.4.3.1 Calibration and Inaccuracy

Using a temperature-controlled oven (Vötsch VT7004), 19 chips from the same batch were characterized from −55 °C to 125 °C (in 10 °C steps) in ceramic DIL packages. The reference sensor was a calibrated Pt-100 resistor temperature. To minimize the effects of oven drift, both the Pt-100 and the chips were placed inside a cavity in a large block of aluminum.

Figure 4.18 shows the sensors' output versus temperature. The opposite trend compared to that of the first WhB implementation is simply due to an inverted BS output. Due to the spread in R_p and R_n, its sensitivity is about 16% less than that in the TT corner. Over temperature, the output of the zoom ADC varies from about 0 to 3.2 over temperature, which is still within its designed full-scale range of −4 to 4.

An individual first-order fit is applied to remove process spread, that is, the spread of $R_{DAC}(T_0)/R_p(T_0)$ and $R_{DAC}(T_0)/R_n(T_0)$ in eq. (4.2). The residual error is then determined by the term $f_{pn}(T - T_0)$ in equation (4.2), which turns out to be quite systematic. Despite the reduction of bridge sensitivity due to process spread, the residual error agrees well with simulations made in the TT corner (maximum error < 0.3 °C). As shown before, this error can then be removed by a fixed polynomial.

Without segment averaging, the 3σ inaccuracy is 0.2 °C after the systematic non-linearity is removed by a fixed fifth-order polynomial (Fig. 4.19a). As discussed in Sect. 4.4.2.3, the jumps around −35 °C, 5 °C, and 55 °C (when the fine code $\mu \approx \pm 0.5$) are caused by the nonlinearity of the first stage. With segment averaging

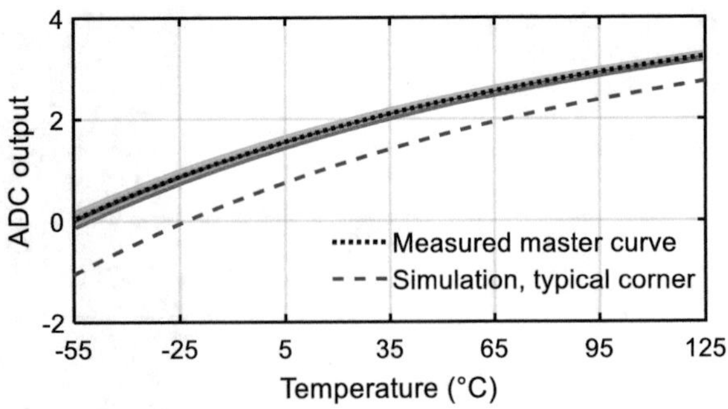

Fig. 4.18 Measured and simulated sensor output versus temperature

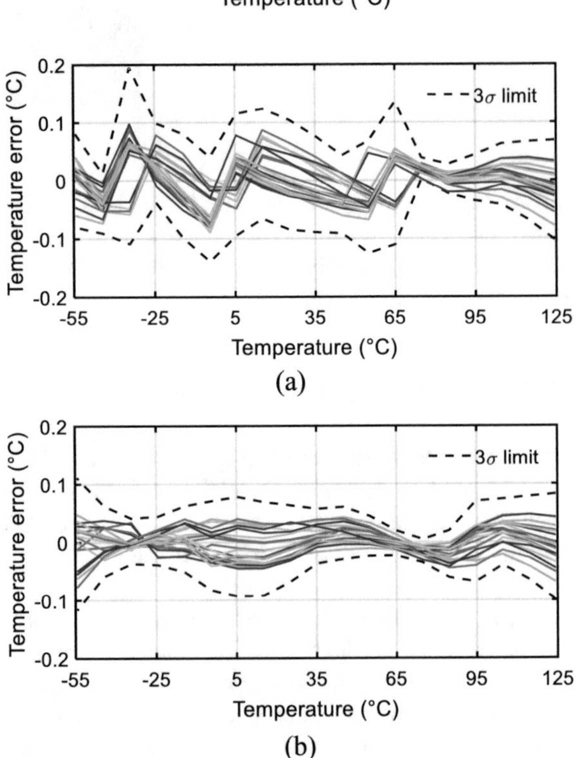

Fig. 4.19 Temperature error after individual first-order fit and systematic nonlinearity removal (**a**) without segment averaging and (**b**) with segment averaging

enabled (Threshold = 0.05), the inaccuracy can be reduced to 0.12 °C (3σ) within the military temperature range (Fig. 4.19b). The 1.6× improvement in accuracy is less than the 8× factor shown in Fig. 4.14, indicating that the majority of the error is due to the spread of the sensing resistors rather than to the nonlinearity of the ADC.

4.4.3.2 Resolution and FoM

With different DEM algorithms, the power spectral densities of the sensor's output bitstream are shown in Fig. 4.20. Since the ADC output range is [−4, 4], 0 dB on the y-axis now corresponds to 1/8 of the full ADC range. Compared to barrel-shifting

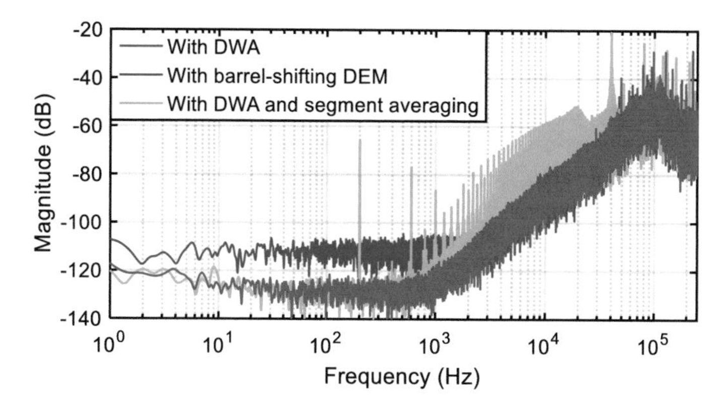

Fig. 4.20 Power spectral density of the bitstream output

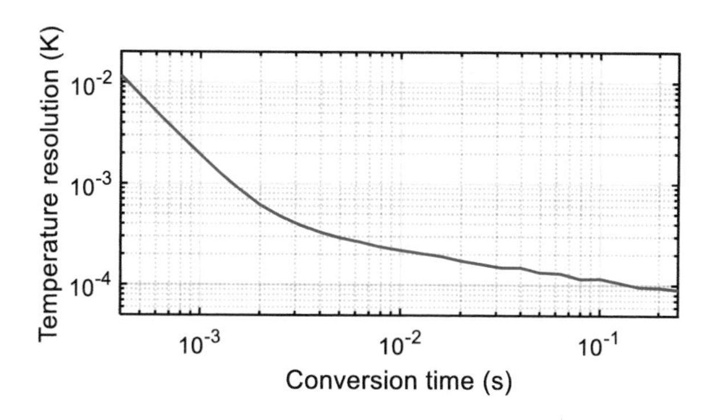

Fig. 4.21 Temperature resolution versus conversion time

DEM, DWA is more complex, but it preserves the sensor's noise floor. Applying segment averaging of 2.5 ms/segment results in tones at multiples of 200 Hz, but not a raised noise floor. For a fixed conversion time of 5 ms (Nyquist frequency of 100 Hz), the tones will be located at the notches of the sinc^2 decimation filter, and thus have no effect on the sensor's resolution. The 1/*f* noise corner is at about 20 Hz, which is mainly due to the nonsilicided poly resistor.

After decimation and drift compensation, the sensor's resolution is derived by calculating the standard deviation in a 1 s interval (Chap. 3, Sect. 3.3.2.1). Its 290 μK_{rms} resolution in 5 ms corresponds to a 40 fJ·K^2 resolution FoM, as shown in Fig. 4.21.

4.4.3.3 Comparison to Implementation I

Compared to the first Wheatstone bridge sensor prototype, this design achieves a 40% larger temperature range and a 1.6× better energy efficiency, while occupying 3× less chip area.

4.5 Implementation III, Even Smaller Area and Better FoM[3]

In this third prototype, an FIR-DAC CTΔΣ-ADC is implemented to replace the zoom ADC used in the second Wheatstone bridge sensor design. The sensor achieves a better resolution FoM of 20 fJ·K^2 with a smaller chip area of 0.12 mm^2 in the same 0.18 μm CMOS process.

4.5.1 *System-Level Design*

As shown in the previous subsection, the use of a multi-bit DAC reduces the swing of the loop filter's input current I_{err}, which decreases the first integrator's power dissipation and the area of C_{int}. In the second prototype, a zoom ADC is used, and the DAC state is determined by combining the result of an initial coarse SAR conversion with the output of a 1-bit CTΔΣM. Extra logic is then required to implement the SAR conversion, as well as the data weighted averaging (DWA) and segment averaging schemes used to mitigate DAC mismatch and amplifier nonlinearity, respectively.

This design employs a 1-bit CTΔΣM with an FIR-DAC, thus ensuring 1-bit linearity without the need for extra logic [6]. For the same DAC resolution, the resulting I_{err} swing will then be about 2× less than that in the zoom ADC, since the FIR-DAC does not require over-ranging (Fig. 4.22). As a result, both the size and area of C_{int} can be reduced, as well as the first integrator's power dissipation. In contrast to a zoom ADC, however, the output of an FIR-DAC may sometimes switch between three DAC levels instead of two. As a result, a more linear input stage is required to prevent quantization noise fold-back, which raises the noise floor [9].

Due to the uncorrelated spread of R_p and R_n, the nominal range of the WhB's output current I_{sig} will spread significantly from batch to batch. To compensate for this, and so make optimal use of the modulator's input dynamic range, a 4-bit batch trim is applied to R_p, so that R_{DAC} is only required to compensate I_{sig} over temperature. As shown in Fig. 4.23, the trimming scheme ensures that only one switch is in series with the selected segment of R_p. Compared to the trimming scheme in [5], this minimizes temperature-sensing errors due to the switches' finite on-resistance.

4.5.2 *Circuit Implementation*

This sensor is designed to operate over the military temperature range from −55 °C to 125 °C. With R_p = 105 kΩ and R_n = 100 kΩ, a 2-bit FIR-DAC consists of four 720 kΩ unit elements results in a BS average ranging from −0.68 to 0.68 at the typical corner. With the 4-bit trimming on R_p (~5.7 kΩ/step) to compensate for process

[3] Pan and Makinwa [23].

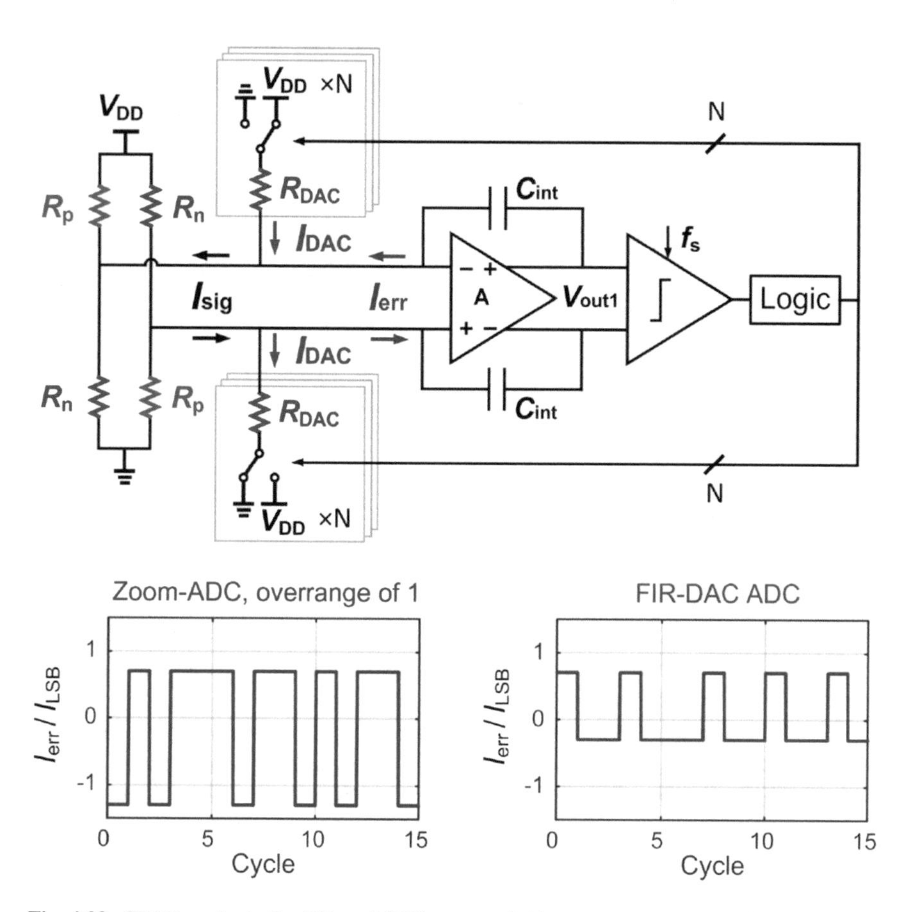

Fig. 4.22 CTΔΣ readout of a differential Wheatstone bridge temperature sensor (top), error current I_{err} over time (bottom)

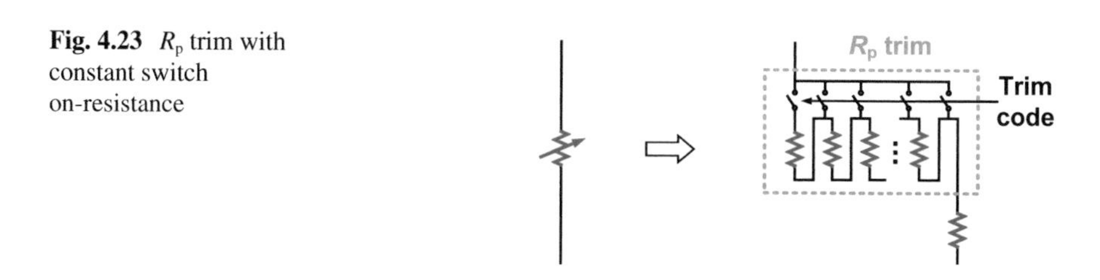

Fig. 4.23 R_p trim with constant switch on-resistance

spread, the BS boundary is then ±0.7. To save area, the DAC resistors are minimum width. Compared to the zoom-based sensor with a similar R_p and the same temperature range, the total RDAC area is reduced by 2.7 × .

To achieve sub-mK resolution in a short conversion time (T_{conv} = 10 ms), the modulator employs a second-order feed-forward architecture (Fig. 4.24). This also reduces the swing at the output of the first integrator, and thus reduces the area of

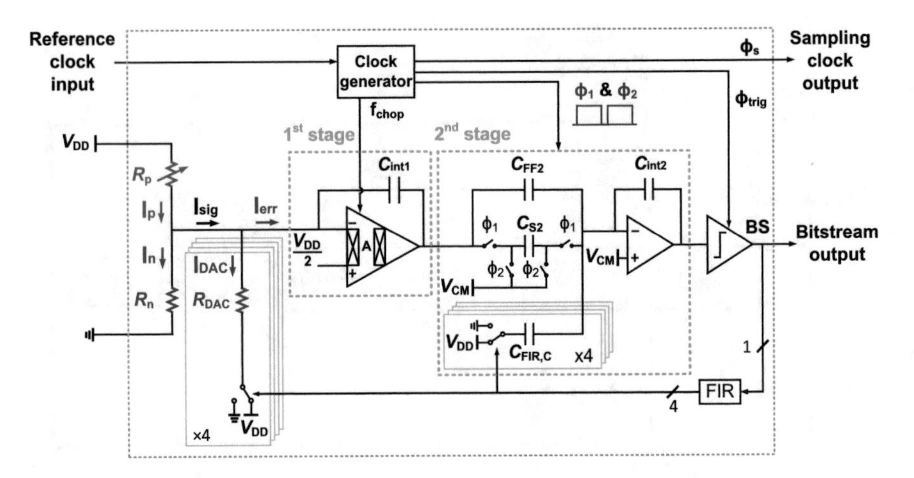

Fig. 4.24 Simplified single-ended system block diagram

C_{int1} (27 pF). The second stage consists of a switched-capacitor integrator (C_{S2} = 100 fF and C_{int2} = 2 pF) and a feedforward path (C_{FF2} = 400 fF), making it a hybrid ΔΣM [15]. An extra FIR filter ($C_{FIR,C}$) is used to compensate for the delay introduced by the FIR-DAC [6].

The first integrator's opamp is a scaled version of that used in the second WB sensor prototype (Fig. 3.30). Compared to an OTA, it has a reduced input swing, and thus exhibits better linearity. It consists of two current-reuse stages which maximize the noise efficiency of the input stage, and, compared to the use of two common-source amplifiers, halves the output stage's bias current for a given maximum output current. Also, the use of high threshold devices enlarges the output swing (~1.5 V at room temperature). This, in turn, allows the area of C_{int1} to be further reduced. The input stage is chopped to suppress its offset and $1/f$ noise, and so the chopping frequency f_{chop} must be chosen such that quantization noise is not down-converted to DC. In this design, f_{chop} can be set to either f_s = 500 kHz or, by exploiting the FIR-DAC's spectral nulls, to multiples of f_s /8 [6], as shown in Fig. 4.25. From simulations, the chopped opamp has a residual $1/f$ corner frequency of 2 Hz, a DC gain of 80 dB, while its input/output stages consume 15 μW/11 μW, respectively. The second stage is built around a cascoded telescopic OTA, which also has an 80 dB gain, but consumes only 3.5 μW.

4.5.3 Measurement Results

Two identical sensors were fabricated on the same die in a 0.18 μm CMOS process (Fig. 4.26). This allows their resolution to be accurately estimated via differential measurements, which effectively reject ambient temperature drift. Each sensor consumes about 44 μA (41 μA analog and 3 μA digital) from a 1.8 V supply, and

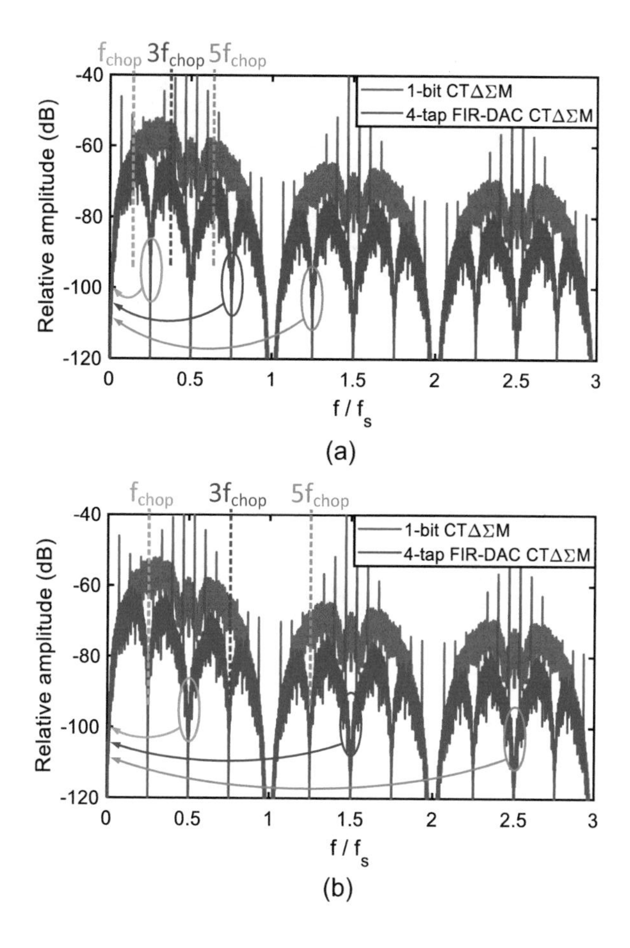

Fig. 4.25 PSD of the signal at the input of the first integrator with a 1-bit DAC or with a 4-tap FIR-DAC. Nulls in the PSD of the FIR-DAC minimize the folded noise when (**a**) $f_{chop} = f_s / 8$ and (**b**) $f_{chop} = 2 f_s / 8$

occupies 0.12 mm^2, 60% of which is occupied by the WhB and the DAC. The sensors share the same clock generation circuit (0.003 mm^2). To further conserve area, C_{int1} (27 pF, MIM) is located directly above the WhB. For flexibility, the sinc2 decimation filters are implemented off-chip.

4.5.3.1 Calibration and Inaccuracy

Twenty samples from one wafer (40 sensors) were mounted in ceramic DIL packages and characterized in a temperature-controlled oven. The packages were mounted in good thermal contact with a large aluminum block. After a batch trim, the residual spread from sample to sample is less than ±3% full scale at RT. To mimic the effect of batch-to-batch spread, the sensor's output was characterized over temperature for two different trim code settings (Fig. 4.27). After a first-order fit to compensate for process spread, the resulting systematic nonlinearity differs by

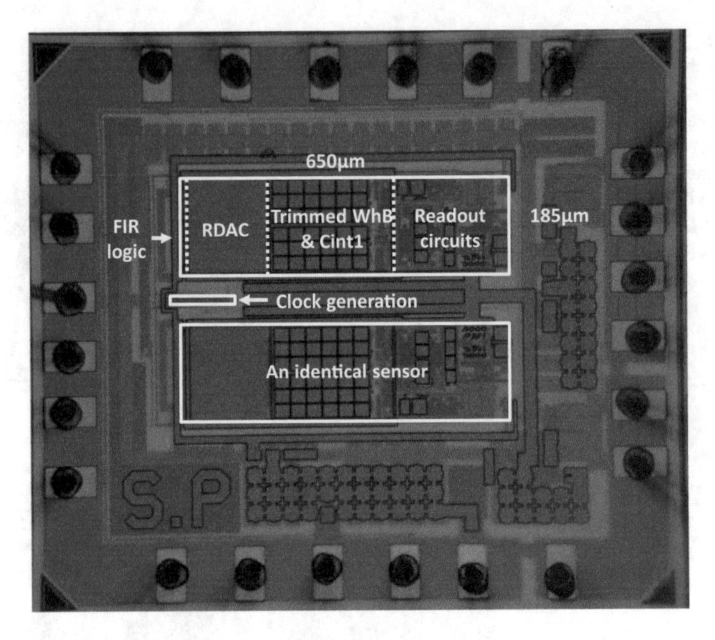

Fig. 4.26 Die micrograph of the fabricated temperature sensor

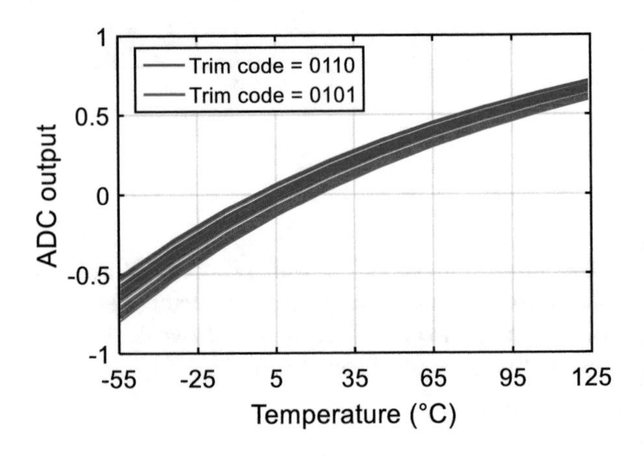

Fig. 4.27 Sensor characteristic with different trimming codes

less than 3 mK for the two trim code settings. It also agrees well with simulations (less than 0.1 °C difference) and so can be robustly corrected by a fixed fifth-order polynomial. The sensor then achieves an inaccuracy of 0.14 °C (3σ) from −55 °C to 125 °C for both trim code settings (Fig. 4.28). The sensor also has a low supply sensitivity: ~0.03 °C/V from 1.6 to 2 V at RT.

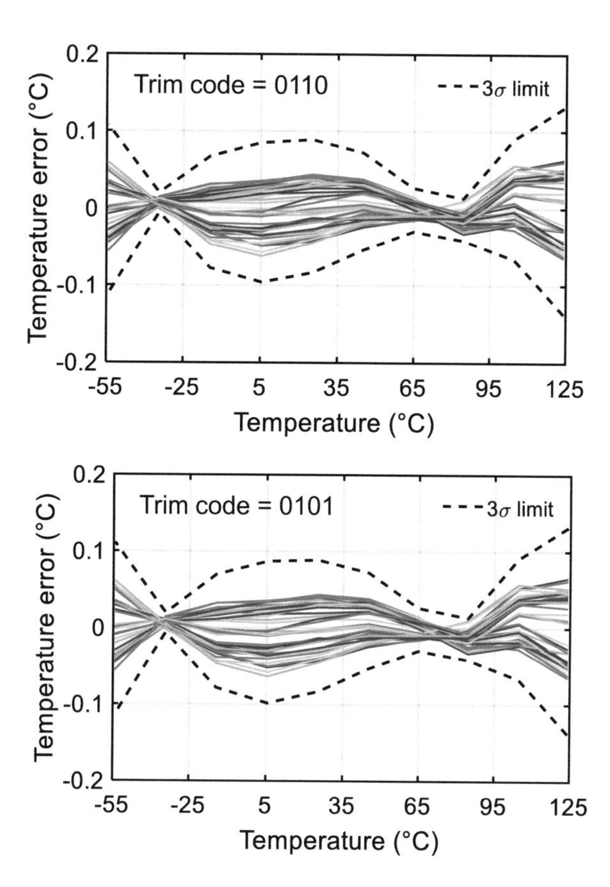

Fig. 4.28 Sensor's temperature error after individual first-order fit and fixed nonlinearity removal with different trimming codes

4.5.3.2 Resolution and FoM

FFTs of the sensor's bitstream output are shown in Fig. 4.29. As designed, the sensor's noise is dominated by the WhB, and changing f_{chop} from f_s to $f_s/8$ has no significant effect on its resolution. The observed $1/f$ noise (~20 Hz corner frequency) is mainly due to the WhB's nonsilicided poly resistors. The effect of oven drift can be suppressed by computing the standard deviation from the difference in the output of the two sensors on each die. As shown in Fig. 4.30, over a 1 s interval, the results of the single-ended and differential approaches agree well, resulting in a resolution of 160 μK (rms) for t_{conv} = 10 ms. Like that in Chap. 3, Sect. 3.4.2.1, the effect of the sensor's own $1/f$ noise can be clearly seen by comparing the resolution derived from different time intervals.

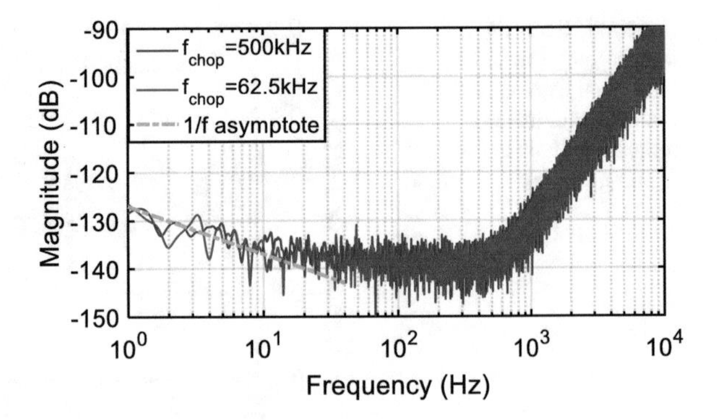

Fig. 4.29 PSD of the sensor's bitstream with different chopping frequencies

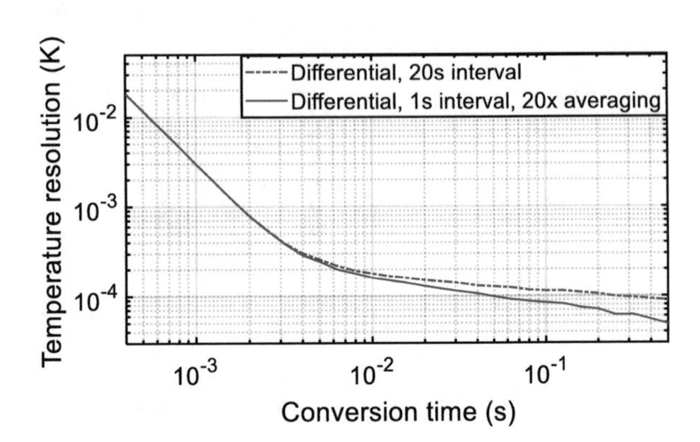

Fig. 4.30 Resolution versus conversion time with different time intervals

4.5.3.3 Comparison to Implementation II

Compared to the second prototype, this design achieves a 2× improvement in both chip area and resolution FoM, while the inaccuracy is kept almost the same.

4.6 Implementation IV, Approaching the FoM Limit[4]

In this design, we target at approaching the practical FoM limit of Wheatstone bridge sensors. This is done by carefully examining all the power/noise contributors and optimizing the readout circuit. This sensor achieves a resolution FoM of 10 fJ·K^2, which is only 6× and 1.5× worse compared to the theoretical and practical FoM limitation, respectively.

[4] Pan and Makinwa [24].

4.6.1 Architecture and Design Considerations

As in the third implementation, this design is also based on the direct digitization of the current output of a Wheatstone bridge by an FIR-DAC CTΔΣ-ADC. To make the best use of ADC range, R_p (~105 kΩ) is 3-bit trimmed to compensate for process spread (~40%), so that R_{DAC} is only required to compensate for the temperature dependence of I_{sig} over the targeted temperature range (−55 °C to 125 °C). To minimize the error from trimming switches, a unary trim is applied to ensure that only one switch is in series with the selected segment of R_p, as shown in Fig. 4.23.

4.6.1.1 RDAC Switching Scheme

As shown in Fig. 4.31a, in previous designs, the DAC resistors were switched between the supply rails. Some of the R_n-type DAC resistors (R_{DAC1}) will then be connected in parallel with the R_p arms while the rest (R_{DAC2}) are connected in parallel with the R_n arms. In a current readout scheme, the addition of balanced DAC

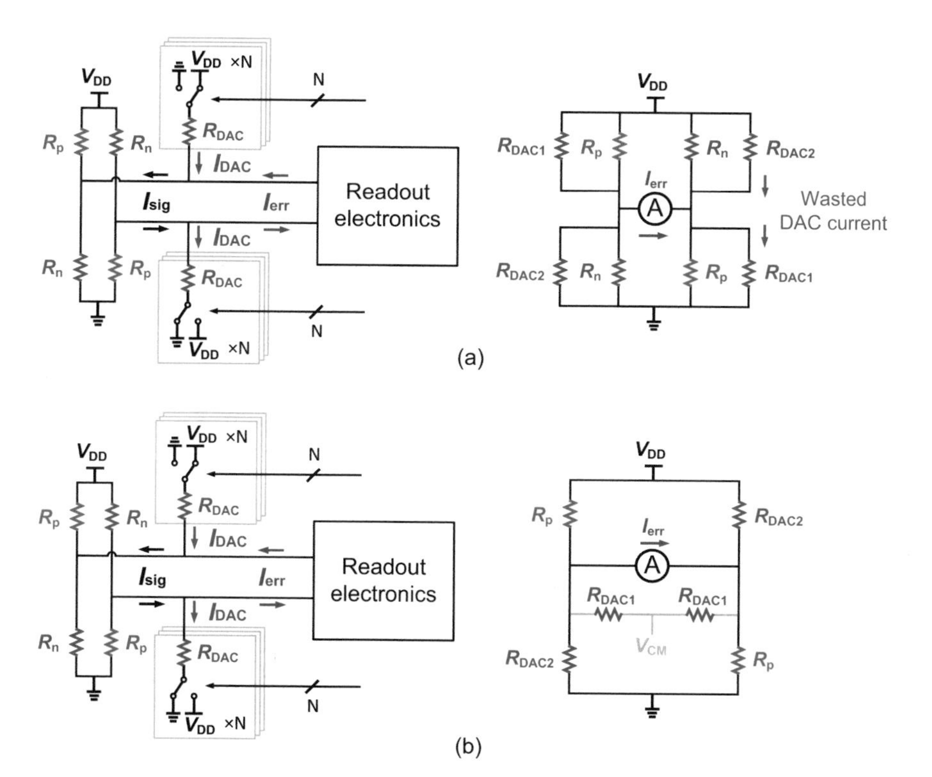

Fig. 4.31 (a) Rail-to-rail DAC switching scheme of a Wheatstone bridge sensor, showing how the DAC resistors consume extra supply current. (b) Proposed return-to-CM RDAC made from unit elements

resistors ($R_{DAC1} = R_{DAC2}$) will not alter the sensitivity of I_{err} to temperature. However, the added noise and power will degrade the FoM of the resistive front-end (WhB and DAC resistors). For the resistive front-end in the third prototype, the FoM will be degraded from 1.7 fJ·K^2 to 3.0 fJ·K^2 at room temperature (RT).

It is worth noting that, to avoid the extra supply current, a 1-bit serial DAC can be realized by using a switch to short a small segment of the R_n branch [16, 17]. However, realizing a linear multi-bit DAC then requires the implementation of non.

uniform resistive segments, whose matching cannot be improved by DEM. Alternatively, the unused resistors of a parallel DAC can be switched to the output common-mode voltage of the bridge V_{CM} (=V_{DD}/2). As shown in Fig. 4.31b, the required V_{CM} can be realized by simply shorting unused DAC pairs together, thus obviating the need for a dedicated voltage reference. Since the voltage drop across the R_{DAC} elements has now been reduced from V_{DD} to V_{DD}/2, their values must be halved to achieve the same DAC currents. For the same (minimum) resistor width, this return-to-CM (RCM) switching scheme also reduces the DAC area by half.

With an appropriately scaled DAC, the proposed RCM switching scheme improves the FoM of the resistive front-end to about 2.2 fJ·K^2 at RT. Due to the noise of the unused DAC resistors (R_{DAC1}), this is somewhat more than the theoretical FoM of the bridge itself (1.7 fJ·K^2). Although it might be tempting to eliminate their noise contribution by letting the unused resistors float, their parasitic capacitances will then cause slow-settling DAC currents. These, in turn, will cause inter-symbol-interference (ISI) and significantly increase the modulator's in-band noise (IBN).

4.6.1.2 DAC Array and DAC Range Optimization

Rather than implementing the R_n and R_{DAC} arms as separate resistors, they can be implemented as a single array of N unit resistors. To balance the bridge, the modulator must drive the DAC such that, on average, $R_{DAC} = R_p$. If the modulator's bitstream average μ ranges from 0 to N, then μ is given by:

$$\mu = \frac{R_{DAC}(T)}{R_p(T)} = \frac{R_{DAC}(T_0)}{R_p(T_0)} \cdot f(T) \tag{4.3}$$

where T_0 is a reference temperature, R_p (T_0) and R_{DAC} (T_0) are nominal resistances, and $f(T)$ represents the combined temperature dependency of both resistors assuming no TC spread. Compared to Eq. (4.2), the offset trim is eliminated. This means that any spread in the nominal resistances R_p (T_0) and R_{DAC} (T_0) can be corrected by a 1-point trim. In practice, TC spread is present, and so a 2-point trim is required for better accuracy.

In this work, the number of parallel DAC resistors ($N = 6$) is a trade-off between chip area and energy efficiency: increasing N decreases I_{err}, but requires more

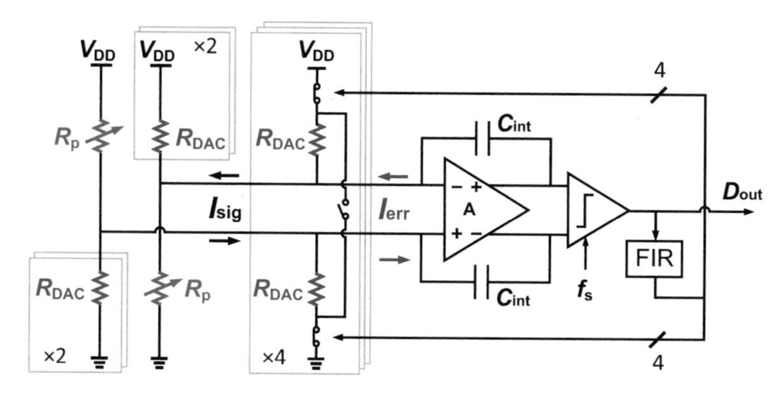

Fig. 4.32 FIR-DAC CTΔΣM readout of the WhB temperature sensor after DAC range optimization

area-consuming DAC resistors. To cover the targeted temperature range, only 4 of the 6 unit elements (each 370 kΩ) are switched, as shown in Fig. 4.32. Thus, the modulator's bitstream average μ ranges from 2 to 6.

4.6.1.3 Integrator Nonlinearity

As introduced in Sect. 4.5.2, one challenge of the multi-bit DAC is integrator nonlinearity and quantization noise fold-back. In the third Wheatstone bridge implementation, the input stage was built around a two-stage opamp, whose high gain keeps its input swing small (<1 mV) and thus mitigates the nonlinearity of its input differential pair. However, the power consumed by its output stage then leads to a loss of energy efficiency.

Greater energy efficiency can be achieved by building the first integrator around a single-stage OTA, since all its supply current then contributes to lowering its noise. However, for the same level of output current I_{out}, this will result in a significantly larger input swing (tens of mVs) and hence nonlinearity [18]. In a conventional differential pair, the maximum I_{out} is limited by the tail current, resulting in a compressing V-I nonlinearity (Fig. 4.33b). In contrast, an expanding and slightly more linear characteristic can be obtained by employing a pseudo-differential (PD) topology (Fig. 4.33a) [19]. In both cases, the effect of OTA nonlinearity can be mitigated by either increasing the tail current or by resistive degeneration, as shown in Fig. 4.33c. However, both approaches reduce energy efficiency. In the next section, a more energy-efficient way of improving OTA linearity will be discussed.

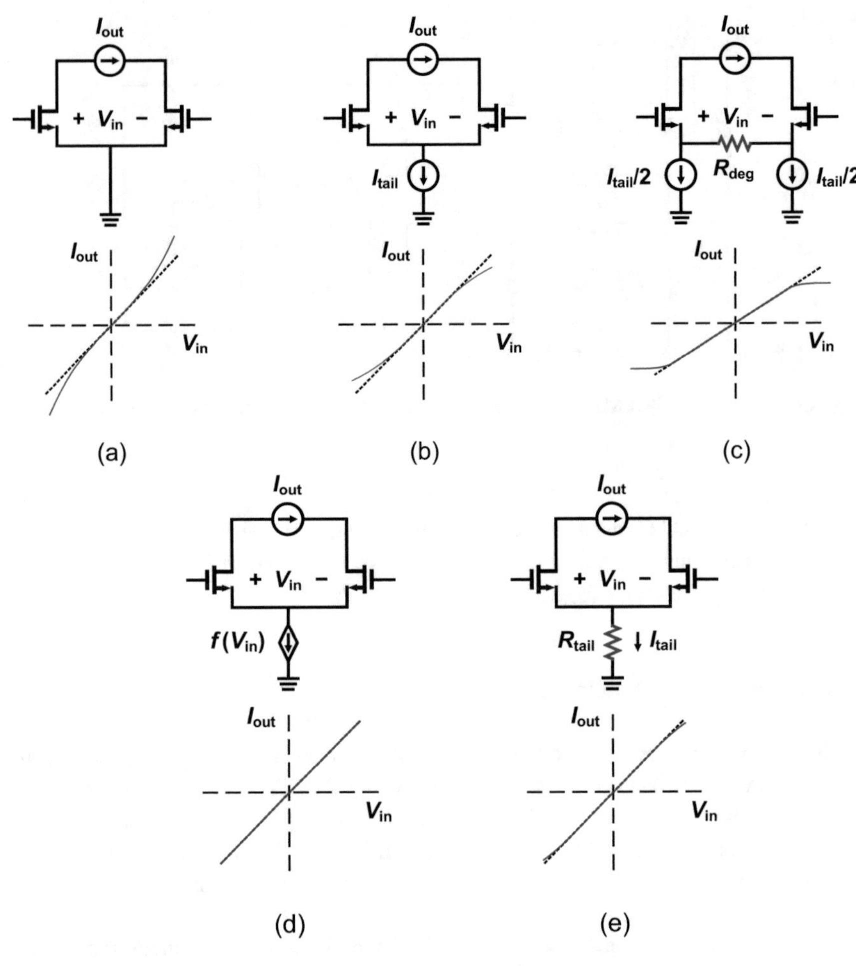

Fig. 4.33 Output characteristics for OTAs with (**a**) zero tail impedance, (**b**) high tail impedance, (**c**) degeneration, (**d**) controlled tail impedance, (**e**) fixed tail impedance

4.6.2 Linearized OTA Design

4.6.2.1 Linearization Principle

From Fig. 4.33a, b, it can be seen that the impedance of the tail current source has a strong influence on OTA nonlinearity. As in [20], it has been shown that the linearity of an OTA can be extended by making its tail current a nonlinear function of its input voltage V_{in} (Fig. 4.33d). Although this approach results in excellent linearity, it requires two trimming knobs to compensate for process spread. As also suggested in a concurrent work [21], a simpler solution is to replace the tail current source by a tail resistor and then optimize its value for linearity, as shown in Fig. 4.33e.

For input transistors biased in weak inversion, which therefore have an exponential I-V characteristic, it can be shown (Appendix A.2) that the value of R_{tail} required to cancel the OTA's dominant third-order nonlinearity is given by:

$$R_{\text{tail}} = \frac{nU_{\text{T}}}{2I_{\text{tail}}}, \tag{4.4}$$

where I_{tail} is the tail current for $V_{\text{in}} = 0$, n is a process-dependent slope factor, and $U_{\text{T}} = kT/q$ is the thermal voltage. It should be noted that this technique also works when the input transistors are biased in moderate inversion, for example, to increase speed. In such cases, however, the required resistor will be somewhat smaller.

4.6.2.2 Biasing Generation

Equation (4.4) indicates that for a fixed R_{tail}, the optimal I_{tail} should be proportional to absolute temperature (PTAT). This can easily be achieved by a conventional constant-gm biasing circuit (Fig. 4.34). Assuming that both M1 and M2 are biased in weak inversion with a current density ratio of k_1:1, and that the current mirror ratio between M3 and M5 is 1:k_2, the output reference current can be expressed as:

$$I_{ref} = \frac{nU_{\text{T}}}{R_{\text{bias}}} \cdot \ln(k_1) \cdot k_2. \tag{4.5}$$

Assuming a ratio of 1:k_3 between I_{ref} and I_{tail}, and combining (4.4) and (4.5) then results in:

$$R_{\text{bias}} = 2R_{\text{tail}} \cdot \ln(k_1) \cdot k_2 \cdot k_3. \tag{4.6}$$

Since R_{bias} is proportional to R_{tail}, a process and temperature robust biasing scheme can thus be achieved by realizing R_{bias} and R_{tail} as a pair of ratiometrically matched resistors.

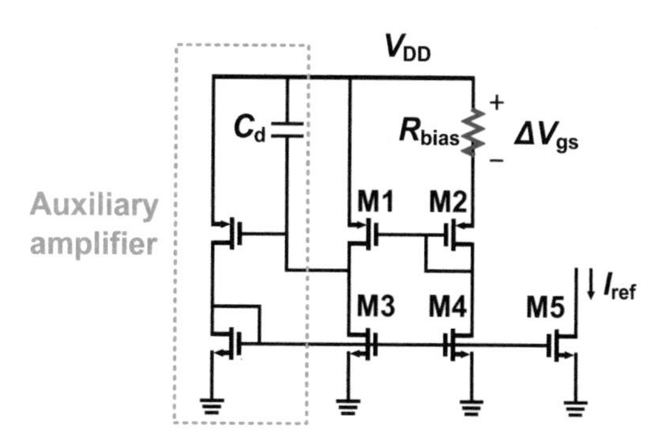

Fig. 4.34 Constant-gm biasing generation

4.6.2.3 Circuit Structure

The simplified schematic of the linearized OTA is shown in Fig. 4.35. To maximize its energy efficiency, its supply current is reused by both PMOS and NMOS input pairs. Both pairs are cascoded to achieve high DC gain. For simplicity, a single PMOS-based constant-Gm biasing circuit was used to set the tail currents of both the NMOS and PMOS input transistors via large biasing resistors R_b. The input voltage is capacitively coupled to the gate of the input transistors. To minimize its noise contribution at the chopping frequency, R_b should be made quite large. This large resistance (~1 GΩ) is achieved by duty-cycling. As in [19], the entire OTA is chopped, allowing it to amplify DC inputs despite its AC-coupled topology. This also suppresses the offset and 1/*f* noise of the OTA. The use of chopping also improves the OTA's low-frequency CMRR and PSRR, thus alleviating one drawback of eliminating the conventional tail current source [21]. A conventional continuous-time CMFB circuit is used. To facilitate experimental comparison, extra switches (not shown) were included to short all the replicas of R_{tail} shown in Fig. 4.35, thus converting the linearized OTA into a PD OTA (Fig. 4.33a) with the same bias current.

4.6.2.4 Nonlinearity Simulation Results

To verify its performance, the linearity of the proposed OTA was simulated and compared to that of conventional and PD OTAs biased at the same tail current. As shown in Fig. 4.36, the use of a tail resistor results in significantly less nonlinearity. Even over process and temperature (−55 °C to 125 °C), the proposed scheme is quite robust. As shown in Fig. 4.37, the worst-case nonlinearity is still 12× better than that of the PD OTA and 40× better than a conventional OTA.

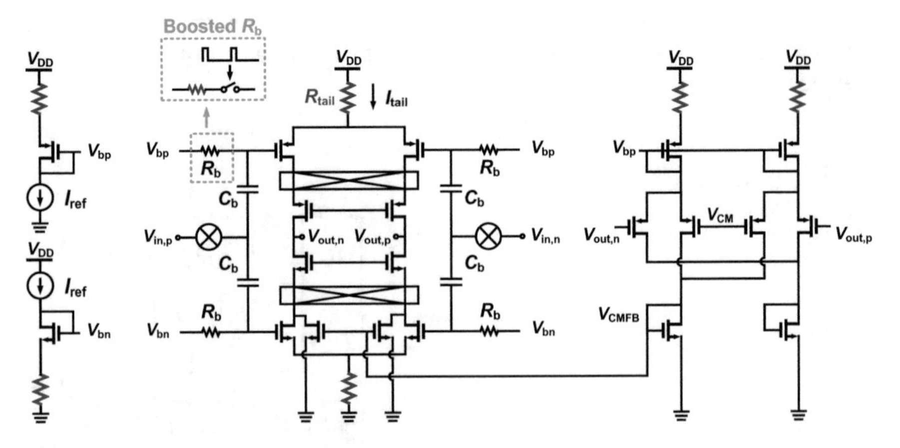

Fig. 4.35 Simplified schematic diagram of the linearized OTA, including the biasing generation and CMFB circuit

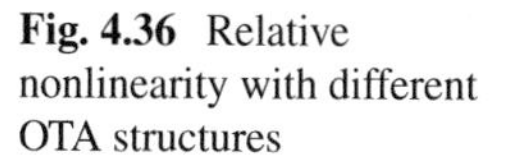
Fig. 4.36 Relative nonlinearity with different OTA structures

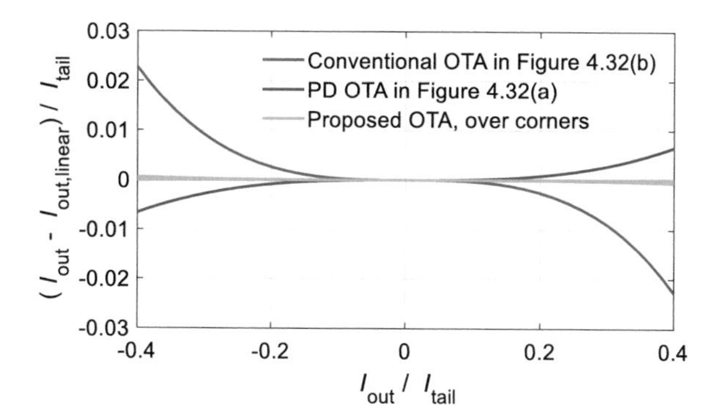

Fig. 4.37 Relative nonlinearity of the proposed OTA over corners

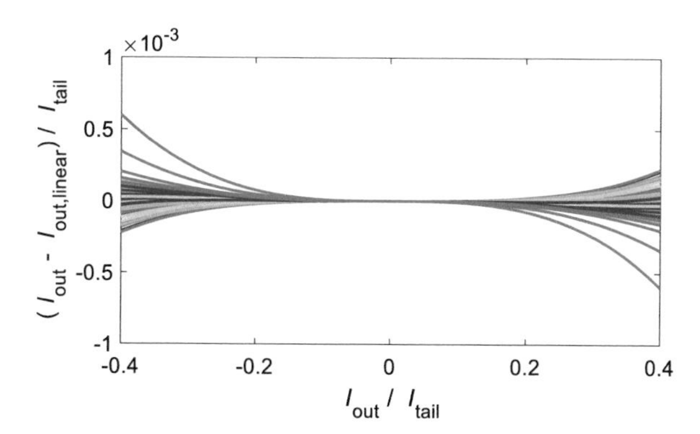

4.6.2.5 Power Scaling and System-Level Simulation

To optimize energy efficiency, the tail current of the OTA used in the first stage of the CTΔΣM was set to 7.6 μA at RT. The first stage then consumes ~9 μA, including the biasing and CMFB circuit, while its input-referred noise is ~20 nV/$\sqrt{\text{Hz}}$. In comparison, the resistive front-end consumes ~17 μA, and its noise level is ~36 nV/$\sqrt{\text{Hz}}$. Neglecting the rest of the CTΔΣM, this results in a theoretical FoM of 4.4 fJ·K^2.

Figure 4.38 shows the results of system-level simulations to verify the effect of OTA nonlinearity on the modulator's in-band noise. With a 4-element FIR-DAC, the maximum swing of I_{err} at RT is about 1/3 I_{tail}. Due to quantization noise folding, the use of a conventional OTA then results in a noise floor that is about 9 dB higher than the expected thermal noise. The improved linearity of a PD OTA reduces quantization noise-folding and brings this to the same level as the thermal noise. After tail-resistor linearization, however, the sensor becomes truly thermal-noise limited, as the folded quantization noise drops to about 20 dB below the thermal noise.

In the third Wheatstone bridge prototype, similar suppression of quantization noise folding is achieved by building the first stage around a two-stage opamp. However, for similar input noise and output current levels, this almost doubles the required supply current (~17 μA including the biasing and CMFB circuits).

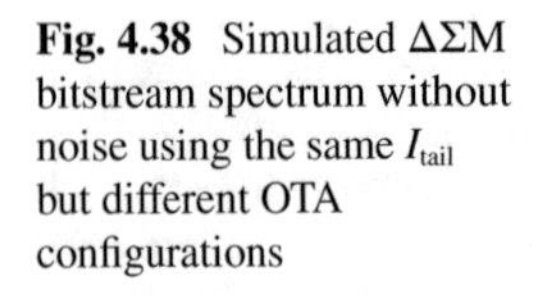

Fig. 4.38 Simulated ΔΣM bitstream spectrum without noise using the same I_{tail} but different OTA configurations

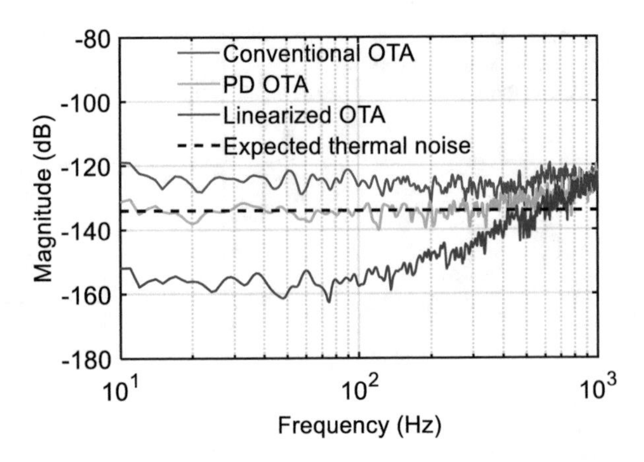

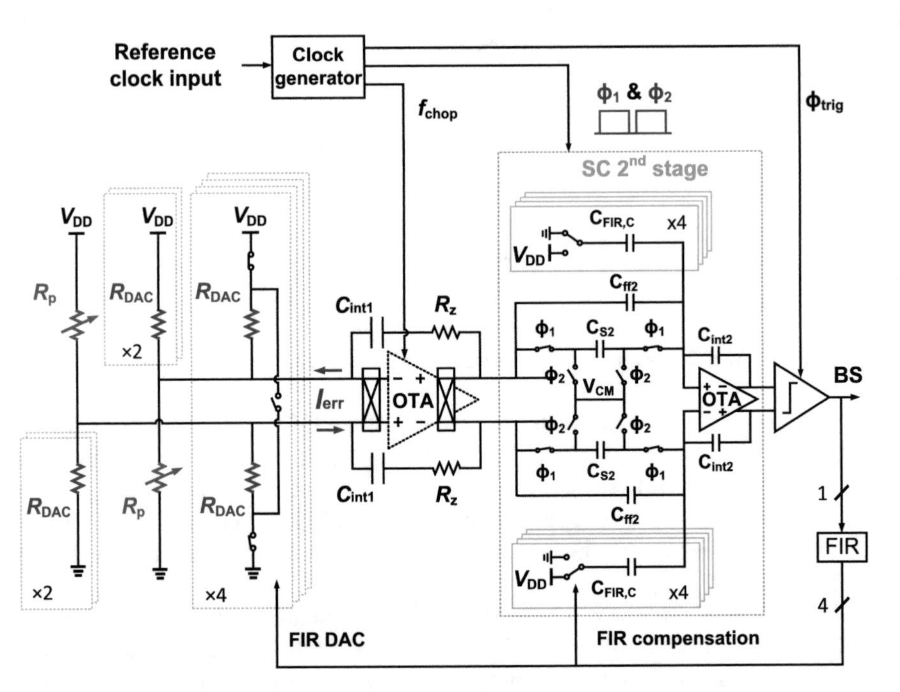

Fig. 4.39 Simplified system block diagram

4.6.3 Circuit Implementation

Figure 4.39 depicts the block diagram of the proposed temperature sensor. As in previous designs, a second-order ΔΣM was adopted to achieve the required resolution in a reasonable conversion time, and a feedforward architecture was chosen to suppress the swing at the output of the first stage, so that the size of C_{int1} (27 pF) can be minimized. To further improve area efficiency, C_{int1} is implemented as a high-density metal-insulator-metal (MIM) capacitor, which is placed above the WhB. The CTΔΣM's sampling frequency (f_s) is set to 500 kHz, which is derived from an off-chip 2 MHz master clock.

The CTΔΣM's first-stage integrator is based on the tail-resistor-linearized OTA introduced in Sect. 4.6.2. A delay-line-based pulse generator operating at $2{\cdot}f_{chop}$ is used to duty-cycle R_b with pulses of ~4 ns. With $C_b \approx 2$ pF and $R_b \approx 700\text{k}\ \Omega$, the occupied chip area is quite small (<4 × 0.002 mm^2). Meanwhile, the R_b noise is heavily filtered by C_b, and is suppressed to ~3 nV/$\sqrt{\text{Hz}}$ at f_{chop}, or ~2% of the OTA's total noise. The OTA achieves an 80 dB gain and a unity-gain bandwidth of 18 MHz with a 1 pF load. To improve its phase response at high frequencies, a zero-cancellation resistor R_z is inserted in series with C_{int}. As shown in Fig. 4.25, the quantization-noise folding can be avoided by choosing $f_{chop} = f_s / 4$.

As in the third Wheatstone bridge prototype, the second stage is implemented as an area-efficient switched-capacitor integrator, with a switched-capacitor FIR-DAC ($C_{FIR,C}$) is inserted to compensate for the delay introduced by the resistive FIR-DAC and stabilize the ΔΣM (Fig. 4.39). Compared to the first stage, the noise and linearity requirements of the modulator's second stage are much more relaxed, so the second stage employs a conventional current-reuse OTA, whose tail current is scaled to 1 μA at RT.

4.6.4 Measurement Results

As shown in Fig. 4.40, four WhB temperature sensors were fabricated on the same die in a standard 0.18 μm process. Two employ a silicided-p-poly/n-poly WhB (s-poly WhB), while the other two employ a silicided-p-diffusion/n-poly WhB (s-diffusion WhB). This is because the silicided diffusion resistor has ~10% higher TC and almost the same voltage dependency compared to the silicided poly resistor. Implementing pairs of sensors on the same die allows ambient temperature drift to be effectively rejected by differential measurements. Sinc^2 filters, implemented off-chip for flexibility, are used to decimate the sensors' bit-stream output.

Each sensor consumes 27.5 μA/3 μA from a 1.8 V analog/ digital power supply, and occupies 0.11 mm^2, of which over 50% is occupied by the WhB and the integration capacitors. Four sensors share two clock generation circuits, each occupying 0.003 mm^2. For supply voltages varying from 1.4 V to 2.0 V, both sensors exhibit a supply sensitivity of about 0.04 °C/V at RT, which is mainly limited by the voltage-dependent R_{on} of the trimming/DAC switches.

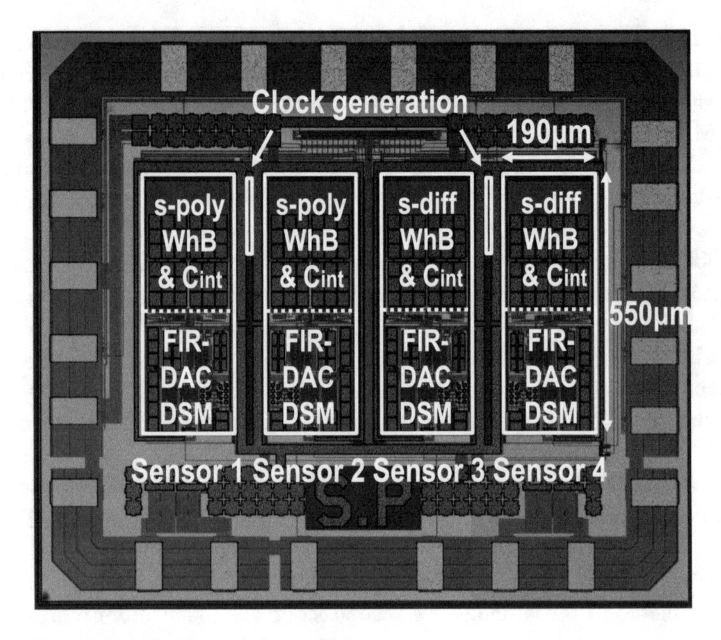

Fig. 4.40 Die micrograph of the fabricated chip

4.6.4.1 Calibration and Inaccuracy

After ceramic DIL packaging, 20 samples from one wafer (i.e., 40 sensors of each WhB type) were characterized in a temperature-controlled oven from −55 °C to 125 °C. To suppress the effects of oven drift, they were mounted in good thermal contact with a large metal block.

The measured performance of the sensors is shown in Fig. 4.40. With the same 3-bit coarse trimming code (011), their spread is about ±3% full scale at RT, or ~7 °C peak-to-peak error. This means that the trimming code can be simply set by measurements on a single sensor. As expected, the s-diffusion WhB has a higher (~6%) sensitivity than the s-poly WhB.

After an individual linear fit to compensate for process spread, the sensors exhibit a systematic nonlinearity. Compared to the second and third Wheatstone bridge implementations, which used the same type of resistors and had the same operating range, the systematic nonlinearity varies by less than 0.1 °C from batch to batch (Fig. 4.41), indicating good repeatability (Fig. 4.42).

As in previous designs, the systematic nonlinearity is removed by a fixed fifth-order polynomial. This results in a 3σ spread of 0.15 °C for the s-poly bridge, and only 0.1 °C for the s-diffusion bridge (Fig. 4.43). Similar results were achieved when the individual linear fit is replaced by a simpler 2-point calibration at −35 °C and 85 °C.

As discussed in Sect. 4.6.1.2, the bitstream average μ is proportional to R_{DAC}/R_p, so that a 1-point gain trim is enough to correct for the spread in their nominal

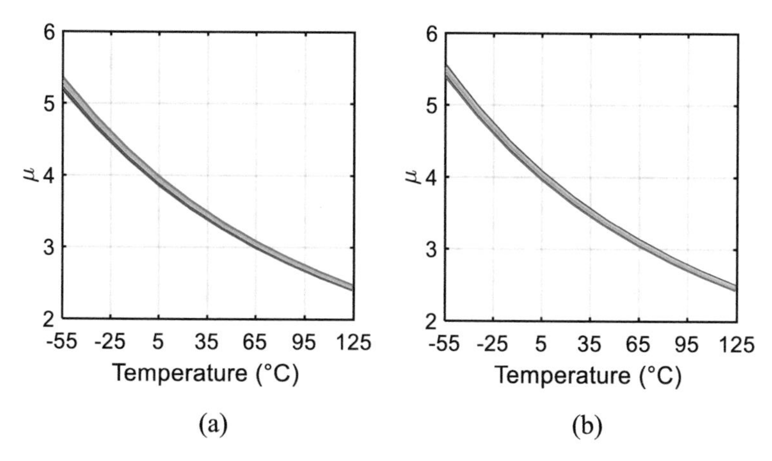

Fig. 4.41 Bitstream average over temperature of (**a**) s-poly WhB sensors and (**b**) s-diffusion WhB sensors before trimming

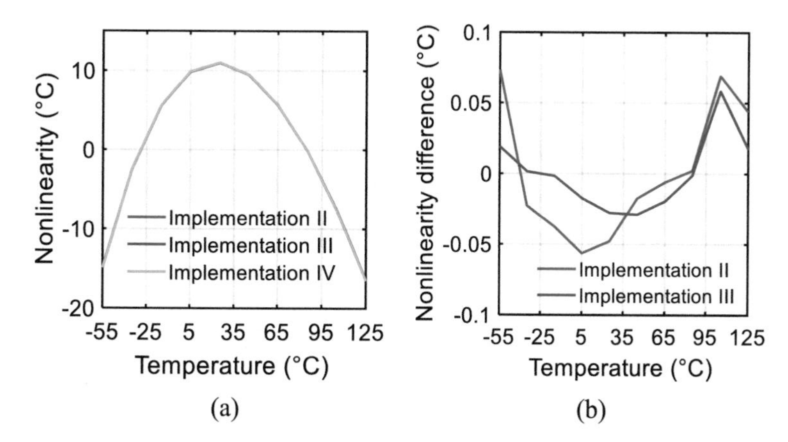

Fig. 4.42 (**a**) Measured systematic nonlinearity of different WhB sensor implementations. (**b**) Difference compared to that of implementation IV

resistances. As shown in Fig. 4.44, doing this results in a residual 3σ spread of 0.65 °C for the s-poly WhB, and 0.4 °C for the s-diffusion WhB.

Alternatively, a correlation-based 1-point trim, like that introduced in Chap. 3, Sect. 3.3.2.2, can be used to exploit the correlation between the sensor's gain error and offset error. Compared to the simple individual gain trim presented here, correlation-based trimming achieves only slightly (~10%) better accuracy, at the expense of the batch calibration needed to determine the correlation coefficients.

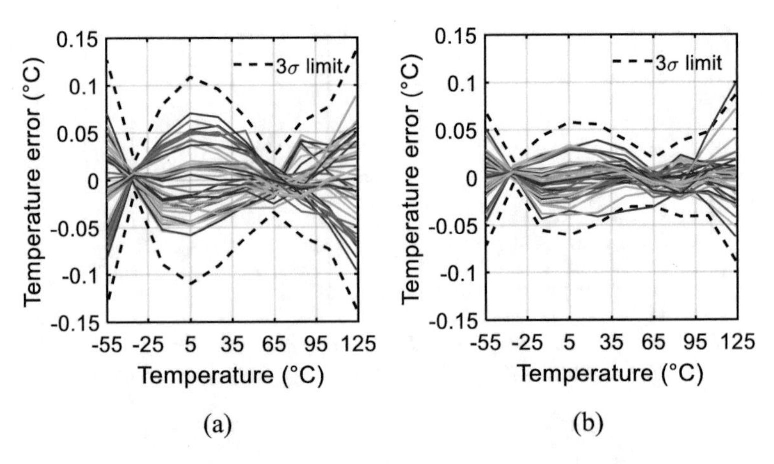

Fig. 4.43 Residual temperature error of (**a**) s-poly WhB sensors and (**b**) s-diffusion WhB sensors after individual first-order fit and systematic nonlinearity removal

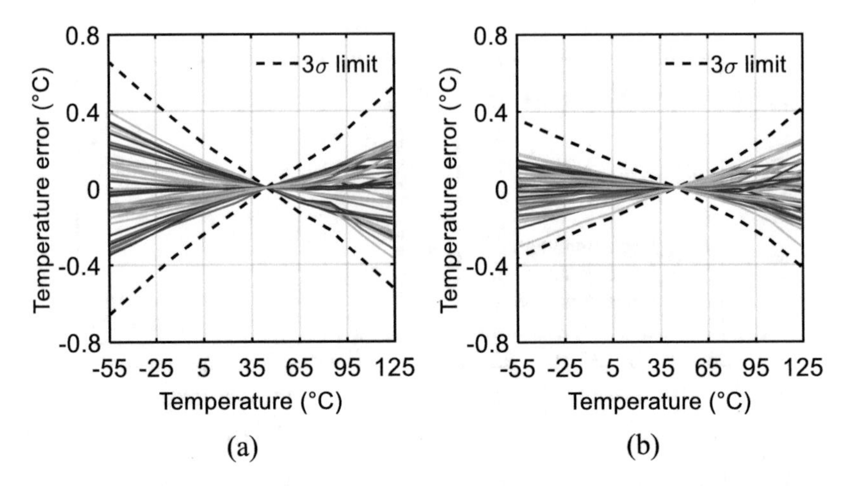

Fig. 4.44 Temperature error of (**a**) s-poly WhB sensors and (**b**) s-diffusion WhB sensors after 1-point trim and systematic error removal

4.6.4.2 Resolution and FoM

Bitstream spectra (20s interval, Hanning window, 10× averaging) of the ΔΣ-ADC's bitstream output are shown in Fig. 4.45, where 0 dB corresponds to an amplitude of 1 in the bitstream average μ (ranges from 2 to 6). Configuring the first-stage OTA as a PD OTA results in a 3-dB increase in the modulator's noise floor, which agrees with the simulation results shown in Fig. 4.38. As in previous designs, the residual $1/f$ noise is mainly due to the non-silicided resistors.

The sensor's resolution is derived via differential measurements, that is, from the standard deviation of the difference in the output of two identical sensors on the

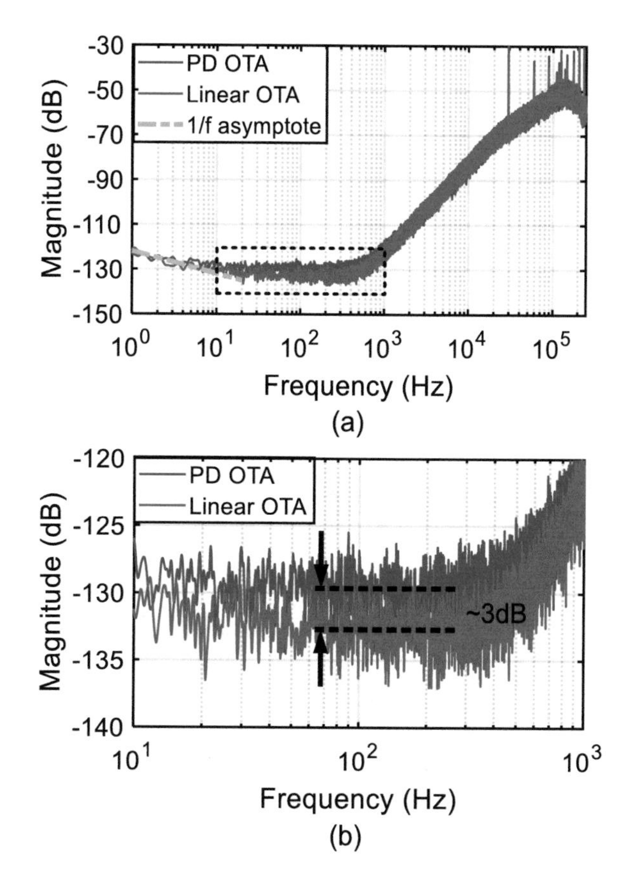

Fig. 4.45 (**a**) Bitstream spectra of the sensor with different OTA configurations (10 M samples, Hanning window, 10× averaging) and (**b**) a zoomed-in plot from 10 Hz to 1 kHz

same die. As shown in Fig. 4.46, with standard deviations computed from bitstream data acquired in a 1 s interval, the s-poly WhB sensor's resolution is estimated to be 160 μK_{rms} in an 8 ms conversion time (T_{conv}). Due to its slightly higher sensitivity, the s-diffusion WhB sensor achieves 150 μK_{rms} in the same T_{conv}. With a 55 μW sensor power, the derived resolution FoMs of s-poly and s-diffusion WhBs are 11 fJ·K^2 and 10 fJ·K^2, respectively.

4.6.4.3 Comparison to Implementation III

Compared to the third prototype, the fourth implementation achieves a 30% better inaccuracy despite occupying less chip area. It also achieves a 2× better energy efficiency, and its resolution FoM (10 fJ·K^2) is close to the practical FoM limit of ~7 fJ·K^2.

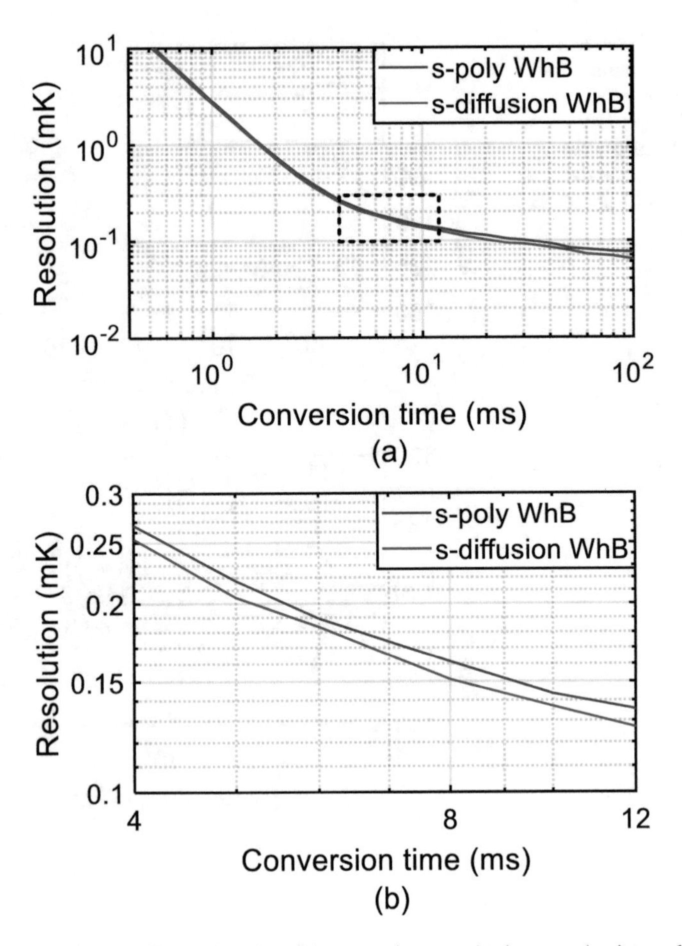

Fig. 4.46 (**a**) Sensor resolution based on bitstream data acquired over a 1 s interval and (**b**) a zoomed-in plot from 4 ms to 12 ms

4.7 Comparison and Concluding Remarks

In this chapter, four Wheatstone-bridge-based temperature sensor prototypes were demonstrated, as summarized in Table 4.1. All the sensors use a second-order CTΔΣ-ADC to directly digitize the temperature-dependent current output of a Wheatstone bridge. By systematically improving each design, their energy efficiency is dramatically enhanced, as can be seen from their FoM: ranging from 65 fJ·K^2 (implementation I) to 10 fJ·K^2 (implementation IV). At the end of this research (2020), implementation IV achieves the best energy efficiency and relative inaccuracy among all Wheatstone bridge (dual-R) temperature sensors.

Table 4.1 Performance summary of the Wheatstone bridge implementations and comparison with the prior art

	JSSC'15 [5]	Implementation I	Implementation II	Implementation III	Implementation IV
Sensor type	Resistor WhB	Resistor WhB	Resistor WhB	Resistor WhB	Resistor WhB
Technology	0.18 μm	0.18 μm	0.18 μm	0.18 μm	0.18 μm
Area [mm^2]	0.43	0.72	0.25	0.12	0.11
Temp. Range [°C]	−40—125	−40—85	−55—125	−55—125	−55—125
3σ Inaccuracy [°C] (trimming points)	0.4[a] (2[b])	0.1 (2[c])	0.12 (2[c])	0.14 (2[c])	0.1 (2[c])
Relative inaccuracy	0.48%	0.16%	0.13%	0.16%	0.11%
Power [μW]	65	180	94	79	55
Conv. time [ms]	0.1	10	5	10	8
Resolution [mK]	10	0.19	0.29	0.16	0.15
Res. FoM [fJ·K^2]	650	65	40	20	10

[a]Min/Max. [b]1-point trim + 1st-order fit. [c]1st-order fit

References

1. J.H. Huijsing, Low noise and low offset operational and instrumentation amplifiers, in *Operational Amplifiers: Theory and Design*, 3rd edn., (Springer, 2017), pp. 307–349
2. R. Wu, J.H. Huijsing, K.A.A. Makinwa, A 21b ±40mV range read-out IC for bridge transducers. IEEE J. Solid-State Circuits **47**(9), 2152–2163 (2012)
3. H. Jiang, S. Nihtianov, K.A.A. Makinwa, An energy-efficient 3.7-nV/$\sqrt{\text{Hz}}$ bridge readout IC with a stable bridge offset compensation scheme. IEEE J. Solid State Circuits **54**(3), 856–864 (2019)
4. H. Jiang, K.A.A. Makinwa, Energy-efficient bridge-to-digital converters, in *Proc. CICC*, (2018), pp. 1–7
5. C.H. Weng, C.K. Wu, T.H. Lin, A CMOS thermistor-embedded continuous-time delta-sigma temperature sensor with a resolution FoM of 0.65 pJ °C^2. IEEE J. Solid State Circuits **50**(11), 2491–2500 (2015)
6. S. Billa, A. Sukumaran, S. Pavan, A 280μW 24kHz-BW 98.5 dB-SNDR chopped single-bit CT ΔΣM achieving <10Hz 1/f noise corner without chopping artifacts, in *ISSCC Dig. Tech. Papers*, (2016), pp. 276–277
7. S. Pan, H. Jiang, K.A.A. Makinwa, A CMOS temperature sensor with a 49fJK^2 resolution FoM, in *Proc. Symp. VLSI Circuits*, (2017), pp. C82–C83
8. S. Pan, Y. Luo, S. Heidary Shalmany, K.A.A. Makinwa, A resistor-based temperature sensor with a 0.13 pJ·K^2 resolution FoM. IEEE J. Solid State Circuits **53**(1), 164–173 (2018)
9. P. Sankar, S. Pavan, Analysis of integrator nonlinearity in a class of continuous-time delta-sigma modulators. IEEE Trans. Circuits Syst. II **54**(12), 661–676 (2007)

10. Y. Chae, K. Souri, K.A.A. Makinwa, A 6.3 μW 20 bit incremental zoom-ADC with 6 ppm INL and 1 μV Offset. IEEE J. Solid State Circuits **48**(12), 3019–3027 (2013)
11. S. Pavan, Efficient simulation of weak nonlinearities in continuous-time oversampling converters. IEEE Trans. Circuits Syst. I **57**(8), 1925–1934 (2010)
12. K. Souri, K.A.A. Makinwa, A 0.12 mm2 7.4 μW micropower temperature sensor with an inaccuracy of ±0.2°C (3σ) from −30°C to 125°C. IEEE J. Solid State Circuits **46**(7), 1693–1700 (2011)
13. B. Yousefzadeh, S.H. Shalmany, K.A.A. Makinwa, A BJT-based temperature-to-digital converter with ±60 mK (3σ) inaccuracy from −55 °C to +125 °C in 0.16 μm CMOS. IEEE J. Solid State Circuits **52**(4), 1044–1052 (2017)
14. S. Pavan, R. Schreier, G.C. Temes, Second-order Delta-Sigma modulation, in *Understanding Delta-sigma Data Converters*, 2nd edn., (Wiley, 2017), pp. 63–82
15. K. Nguyen et al., A 106-dB SNR hybrid oversampling analog-to-digital converter for digital audio. IEEE J. Solid State Circuits **40**(12), 2408–2415 (2005)
16. K.A. Sankaragomathi, J. Koo, R. Ruby, B.P. Otis, A ±3ppm 1.1mW FBAR frequency reference with 750MHz output and 750mV supply, in *IEEE ISSCC Dig. Tech. Papers*, (2015), pp. 454–455
17. S. Hacine, T.E. Khach, F. Mailly, L. Latorre, P. Nouet, A micropower high-resolution ΣΔ CMOS temperature sensor. Proc. IEEE Sensors, 1530–1533 (2011)
18. W. Sansen, Distortion in elementary transistor circuits. IEEE Trans. Circuits Sys. II **46**(3), 315–325 (1999)
19. B. Gönen et al., A continuous-time zoom ADC for low-power audio applications. IEEE J. Solid State Circuits **55**(4), 1023–1031 (2020)
20. R. Sehgal et al., A 13-mW 64-dB SNDR 280-MS/s pipelined ADC using linearized integrating amplifiers. IEEE J. Solid State Circuits **53**(7), 1878–1888 (2018)
21. M.S. Akter, R. Sehgal, K. Bult, A resistive degeneration technique for linearizing open-loop amplifiers. IEEE Trans. Circuits Sys. II **67**(11), 2322–2326 (2020)
22. S. Pan, K.A.A. Makinwa, A 0.25 mm2 resistor-based temperature sensor with an inaccuracy of 0.12 °C (3σ) from −55 °C to 125 °C. IEEE J. Solid-State Circuits **53**(12), 3347–3355 (2018)
23. S. Pan, K.A.A. Makinwa, A Wheatstone bridge temperature sensor with a resolution FoM of 20fJ·K2, in IEEE ISSCC Dig. Tech. Papers., 186–188 (2019)
24. S. Pan, K.A.A. Makinwa, A 10fJ·K2 Wheatstone bridge temperature sensor with a tail-resistor-linearized OTA. IEEE J. Solid-State Circuits **56**(2), 501–510 (2021)

Chapter 5
Application-Driven Designs

5.1 Introduction

In Chaps. 3 and 4, WB- and WhB-based temperature sensors intended for the temperature compensation of MEMS/crystal frequency references were presented. However, this is only one of the many possible applications of resistor-based temperature sensors. In this chapter, two sensors intended for other applications are presented: one intended for biomedical applications, and the other intended for use as a building block in a RC frequency reference.

5.2 A Low-Power Sensor for Biomedical Applications[1]

5.2.1 Background Introduction

In wearable/implantable biomedical applications, body temperature (~37.5 °C) must often be measured accurately, for example, with errors less than 0.1 °C from 39.0 °C to 41.0 °C, and less than 0.2 °C from 35.8 °C to 41.0 °C [1]. This requires temperature sensors with sufficient resolution (<40 mK) in a short conversion time (<100 ms) to facilitate rapid, and thus low cost, calibration. Furthermore, since biomedical devices are typically powered by small thin-film batteries, their sensors should also have high energy efficiency and low-power dissipation. Last but not the least, such sensors should be robust to supply and clock reference variations, as a stable supply/clock is not always available in biomedical environments.

[1] S. Pan and K. A. A. Makinwa, "A 6.6 μW Wheatstone-bridge Temperature Sensor for Biomedical Applications," in *IEEE Solid-State Circuits L.*, vol. 3, pp. 334-337, 2020.

S. Pan, K. A. A. Makinwa, *Resistor-based Temperature Sensors in CMOS Technology*, ACSP · Analog Circuits and Signal Processing,
https://doi.org/10.1007/978-3-030-95284-6_5

BJT- or MOS-based temperature sensors are often used in biomedical applications due to their low-power dissipation (<2 μW) and high resolution [2–4]. Although Wheatstone bridge sensors achieve state-of-the-art energy efficiency (Chap. 4), they typically dissipate more power (>50 μW) and are quite nonlinear, requiring a complex digital backend to perform polynomial linearization. Some resistor-based sensors [5, 6] dissipate much less power (<0.1 μW); however, they are much less energy efficient, and their limited resolution (>300 mK) makes them unsuitable for biomedical applications.

This subsection describes a low-power Wheatstone bridge sensor that meets biomedical requirements, while maintaining high energy efficiency. After a PWM-based 1-point trim, it achieves an inaccuracy of +0.2 °C/−0.1 °C (3σ) over a ±10 °C range centered on 37.5 °C. The use of PWM-based trimming obviates the need for a complex digital backend that implements a high-order linearizing polynomial and a correlated gain/offset trim. It is also quite energy efficient, achieving 200 μK resolution in a 40 ms conversion time while dissipating 6.6 μW, which corresponds to a state-of-the-art resolution FoM of 11 fJ·K^2. A power-down mode allows its average power to be significantly reduced, by duty-cycling, to ~700 nW at 10 conversions/s.

5.2.2 *Circuit Implementation*

5.2.2.1 Wheatstone Bridge and Series DAC

To maximize its sensitivity, as in Chap. 4, the Wheatstone bridge (WhB) sensor employs resistors with opposite temperature coefficients (TCs): silicided poly resistors (R_p) and n-poly resistors (R_n). Since the bridge dominates both the sensor's area and power dissipation, there is a trade-off between these two important parameters. With $R_n \approx R_p \approx 600$ kΩ, that is, ~6× more than that in third/fourth WhB sensor implementation in Chap. 4. The bridge consumes ~4.3 μW from a 1.6 V supply and occupies 0.06 mm^2.

In all the designs presented in Chap. 4, the WhB sensor outputs are digitized by a continuous-time delta-sigma modulator (CTΔΣM) that uses a parallel n-poly DAC (R_{nDAC}) to dynamically balance the bridge (Fig. 5.1a). In steady state, the CTΔΣM's bitstream average μ can be expressed as:

$$\mu = R_{nDAC} / R_n - R_{nDAC} / R_p. \tag{5.1}$$

Since R_{nDAC} and R_n are both n-poly resistors, the first term is a constant. However, since the absolute TC of R_p (0.29%/°C) is larger than that of R_n (−0.15%/°C), the second term is proportional to $1/T$ and is thus rather nonlinear. Together with the resistors' higher-order TCs, this results in a nonlinearity of ~0.3 °C over the desired range (±10 °C range around 37.5 °C), which would then necessitate the use of polynomial nonlinearity correction (Chap. 4).

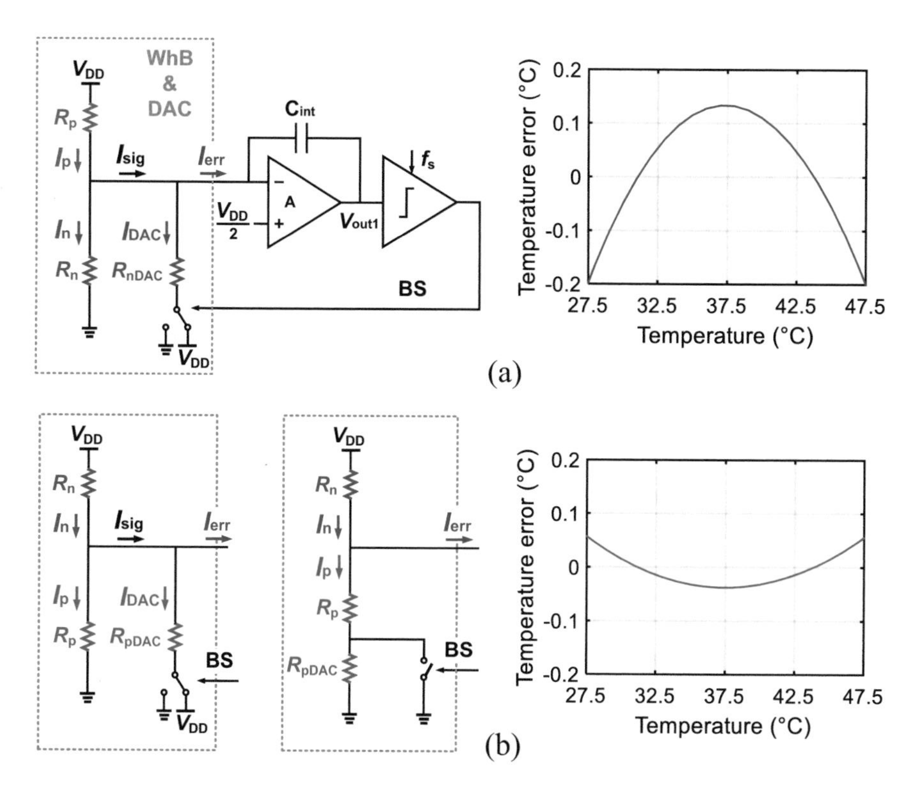

Fig. 5.1 Simulated nonlinearity of a Wheatstone bridge (WhB) sensor with (**a**) a parallel R_n DAC, and (**b**) with a parallel/series R_p DAC

This nonlinearity can be mitigated by realizing the DAC with R_p resistors (Fig. 5.1b), in which case, μ can be expressed as:

$$\mu = R_{\mathrm{pDAC}} / R_{\mathrm{n}} - R_{\mathrm{pDAC}} / R_{\mathrm{p}} \tag{5.2}$$

This reduces the nonlinearity to ~0.1 °C over the same temperature range, thus obviating the need for digital linearization.

Since the desired temperature range is small (20 °C), the resistance change in the Rn branch will also be small, and a large parallel resistor ($R_{\mathrm{pDAC}} \approx 18{\cdot}R_{\mathrm{p}}$) is required to balance the bridge. To save area, a series DAC is used [7], which requires a much smaller ($R_{\mathrm{pDAC}} \approx R_{\mathrm{p}}/18$) resistor (Fig. 5.1b, middle).

In this case, μ can be expressed as:

$$\mu = (2k+1) - (2k+2) \cdot R_{\mathrm{p}} / R_{\mathrm{n}}. \tag{5.3}$$

where $k = R_{\mathrm{p}}/R_{\mathrm{pDAC}}$. As in (5.2), its nonlinearity is also determined by the $R_{\mathrm{pDAC}}/R_{\mathrm{n}}$ and so remains the same.

5.2.2.2 PWM-Assisted Trim

To compensate for process spread, the R_p/R_n ratio can be adjusted by trimming the R_n branch. However, compensating for the worst-case process spread (±40%) and achieving sufficient trimming resolution (<0.05 °C) requires a 12-bit trim DAC. To save area, this is implemented by combining a 5-bit resistor DAC (~15 kΩ/step) with a 7-bit PWM DAC. The former ensures efficient use of the modulator's dynamic range, while the latter provides sufficient resolution. As shown in Fig. 5.2, the resistor DAC is implemented with 31 series resistors, and the PWM trim is implemented by duty-cycling an extra series resistor at $F_{PWM} = F_S/128$, where F_S (=32 kHz) is the sampling frequency of the modulator. To ensure a constant switch on-resistance, a dummy switch controlled by the inverse of the PWM signal (!PWM) is added in series with the duty-cycled resistor. Although quite area efficient, the PWM trim adds a small AC component to the output current of the bridge I_{err}, which uses up ~20% of the modulator's input dynamic range.

5.2.2.3 Return-to-Zero DAC and DSM Readout

One drawback of a series R_p DAC is that it modulates the resistance of the R_p branches. Due to their parasitic capacitances, the resulting rise/fall times of I_{err} (~0.5 μs) are quite significant when compared to the clock period (31.25 μs). Using non-return-to-zero (NRZ) DAC pulses will then cause inter-symbol-interference (ISI), and degrade the modulator's noise floor. To mitigate this, a return-to-zero (RZ) DAC is implemented by splitting R_{pDAC} in two and returning to the resulting middle level after a half clock period, as shown in Fig. 5.3b.

To improve the sensor's supply sensitivity, dummy switches (S_{dummy}) are also inserted in series with the DAC switches (S_1-S_3), so that there are always two series switches in both WhB branches, which cancels the effect of the voltage dependency of switch on-resistance.

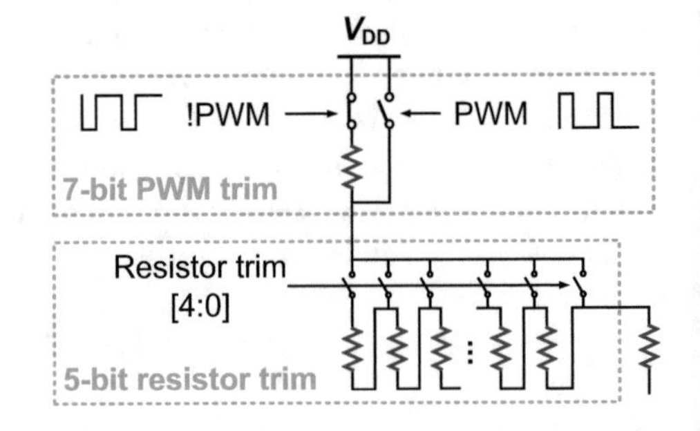

Fig. 5.2 Hybrid R_n trim based on resistor segmentation and PWM

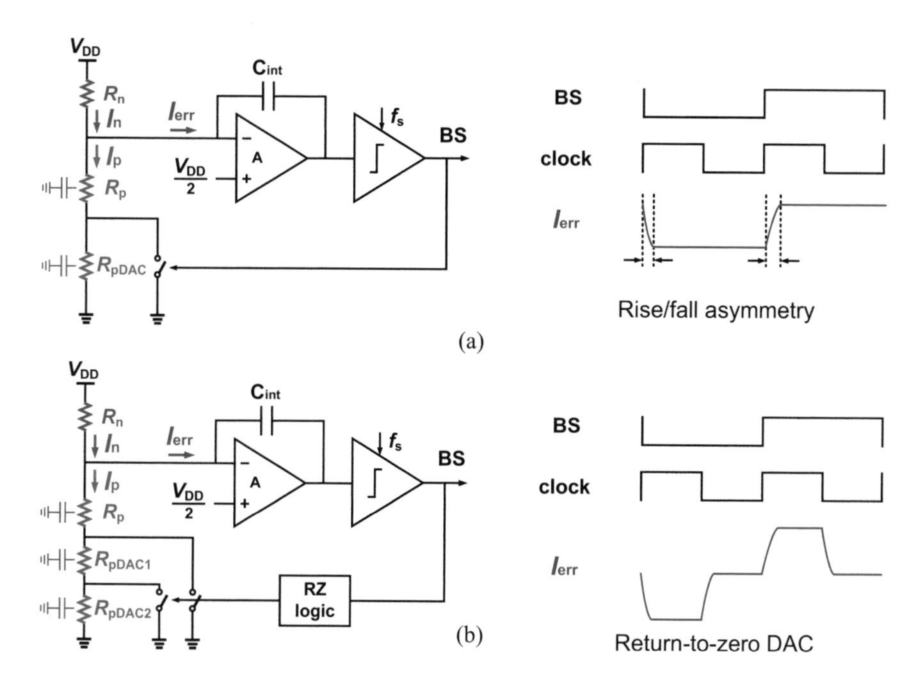

Fig. 5.3 (**a**) Rise/fall asymmetry when using a non-return-to-zero DAC. (**b**) Return-to-zero (RZ) DAC achieved by splitting R_{pDAC}

The sensor's system block diagram is shown in Fig. 5.4a. To achieve high resolution, the bridge is read out by a 2nd-order DSM. As in the 2nd Wheatstone bridge implementation (Chap. 4, Sect. 4.4), the 1st stage is built around a current-reuse OTA, which employs high threshold voltage (high-V_T) input transistors to maximize its output range, and chopping to suppress its $1/f$ noise. The 2nd stage employs another current-reuse OTA and an area-efficient switched-capacitor filter.

Two extra operating modes have also been implemented. First, to test the effectiveness of the RZ DAC, an NRZ DAC can be implemented by modifying the timing of the signals that drive S_1-S_3, as shown in Fig. 5.4b. Second, to implement a power-down mode, the WhB is disconnected from the supply by turning S_{dummy} off. The amplifiers are also switched off by shorting the gates of their PMOS/NMOS tail-current sources to V_{DD}/Gnd. For fast start-up, capacitors pre-charged to Gnd/V_{DD} can be used to restore their gate voltages within one clock cycle [6].

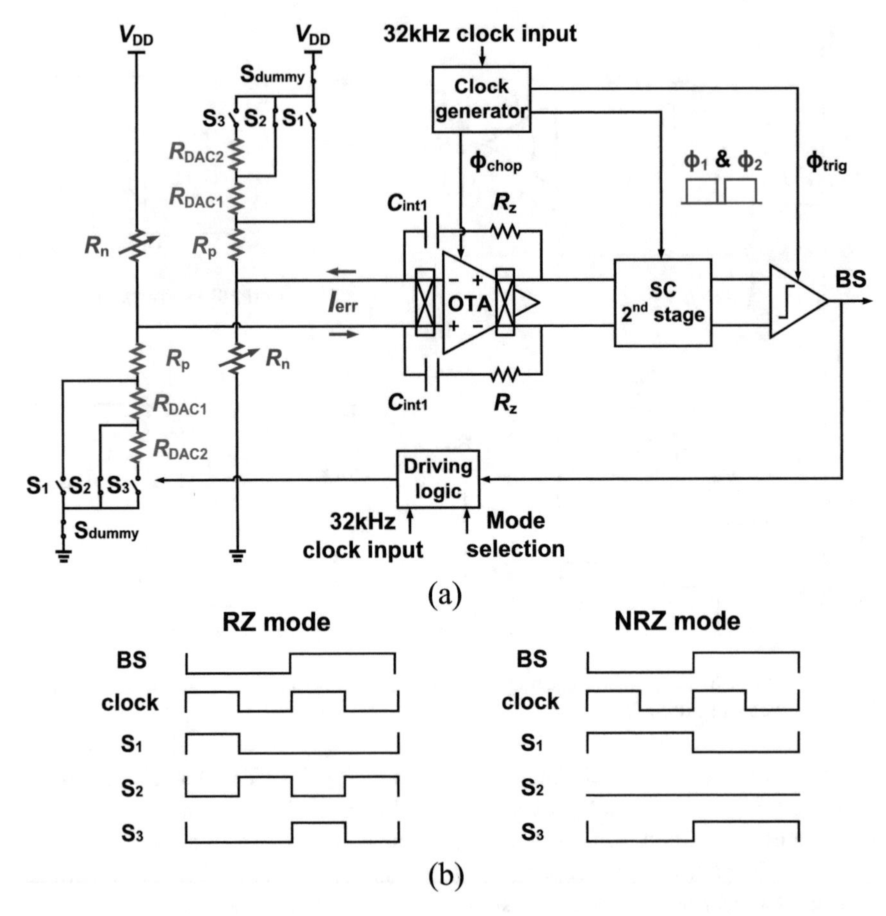

Fig. 5.4 (**a**) Simplified sensor block diagram. (**b**) DAC switching scheme under RZ or NRZ mode

5.2.3 Measurement Results

As shown in Fig. 5.5, two sensors are fabricated on the same die in a standard 0.18 μm CMOS process, allowing ambient temperature drift to be cancelled by differential measurements. They share the same clock/PWM generation circuit (0.003 mm^2), and for flexibility, their decimation filters (sinc^2) are off-chip. The large 1st-stage integration capacitors C_{int1} (MIM, 40 pF) are located above the WhB (0.072 mm^2), while the trimming circuits occupy 0.007 mm^2. At $F_S = 32$ kHz, each sensor consumes 6.6 μW (4.2 μW bridge, 2.1 μW analog, and 0.3 μW digital) from a 1.6 V supply, and occupies 0.12 mm^2.

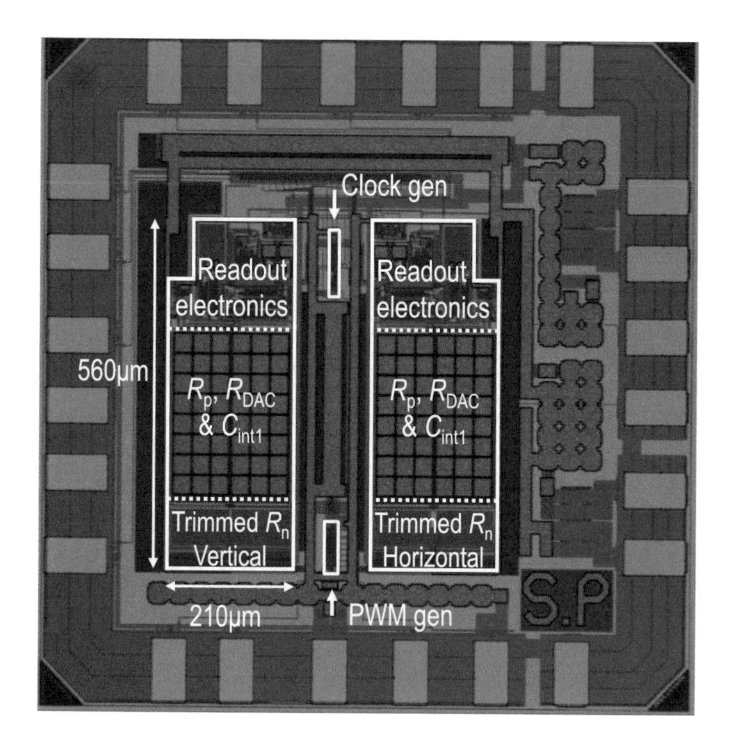

Fig. 5.5 Die photo of the fabricated chip

5.2.3.1 Calibration and Inaccuracy

The sensors (21 ceramic-packaged samples from one wafer) were characterized from 27.5 °C to 47.5 °C in a temperature-controlled oven. Without trimming, the sensor's inaccuracy is about 10 °C. After a 5-bit resistor trim at 37.5 °C, the residual spread relative to a linear master curve is about ±2.5 °C (Fig. 5.6a). This can be reduced to ±0.25 °C (3σ) by applying a digital offset trim (Fig. 5.6b).

As indicated by (5.3), some of the remaining spread is caused by the residual error in the ratio R_p/R_n, which causes sensitivity variations that cannot be eliminated by the aforementioned offset trim. By trimming R_p/R_n (5-bit resistor trim + 7-bit PWM trim), this error is greatly suppressed, as shown in Fig. 5.6c. Despite the residual trimming error at 37.5 °C, the residual spread is then below +0.2 °C/−0.1 °C (3σ) without any additional post-processing (Fig. 5.6d).

To investigate the effects of packaging stress, 14 plastic-packaged chips from the same wafer were also characterized. This causes a systematic sensitivity error and increased spread. Using the linear master curve obtained from the ceramic-packaged chips results in +0.25/−0.1 °C (3σ), as shown in Fig. 5.7.

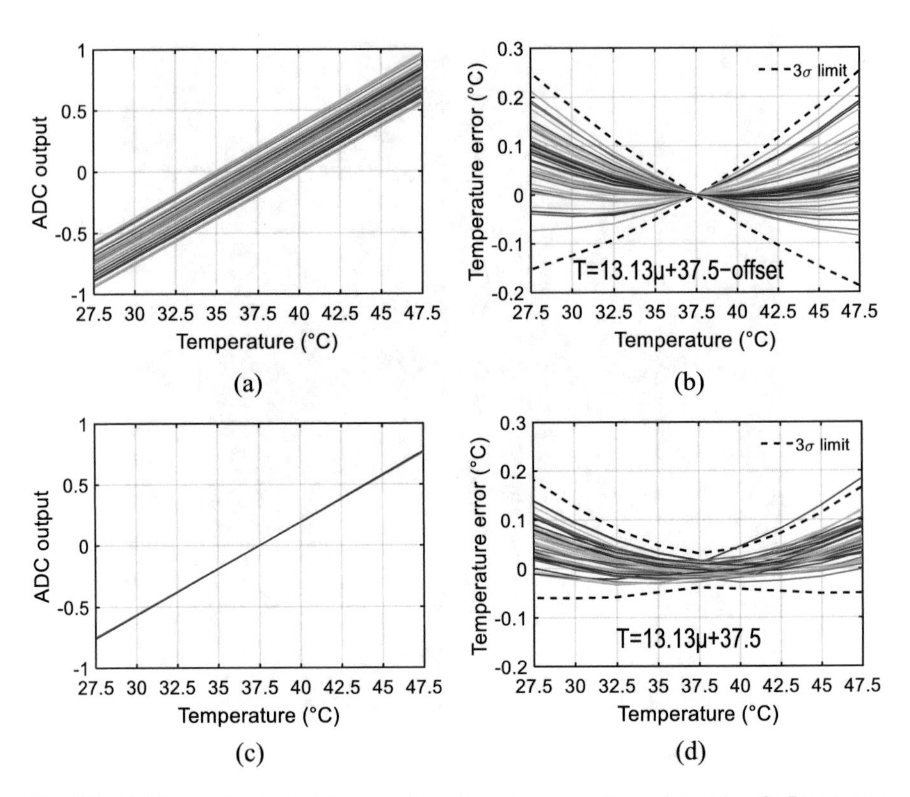

Fig. 5.6 (**a**) Measured output of the ceramic-packaged sensors after resistor trim. (**b**) Inaccuracy after digital offset trim. (**c**) Measured output after PWM trim. (**d**) Inaccuracy after PWM trim without post-processing

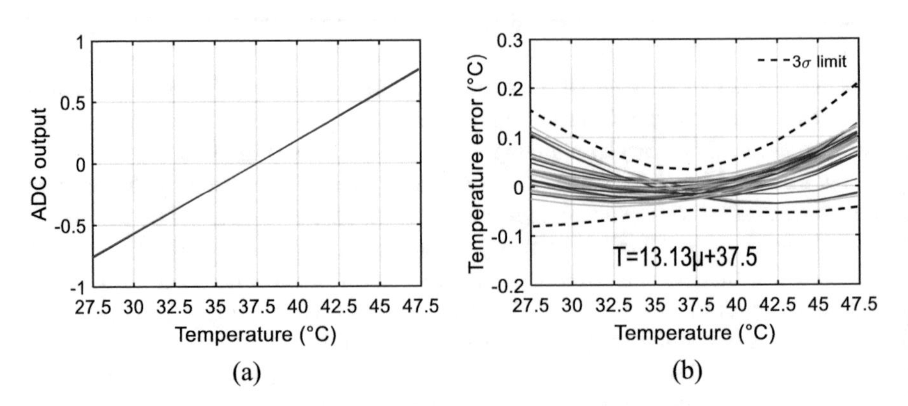

Fig. 5.7 (**a**) Measured output of plastic-packaged sensors after PWM trim. (**b**) Inaccuracy after PWM trim

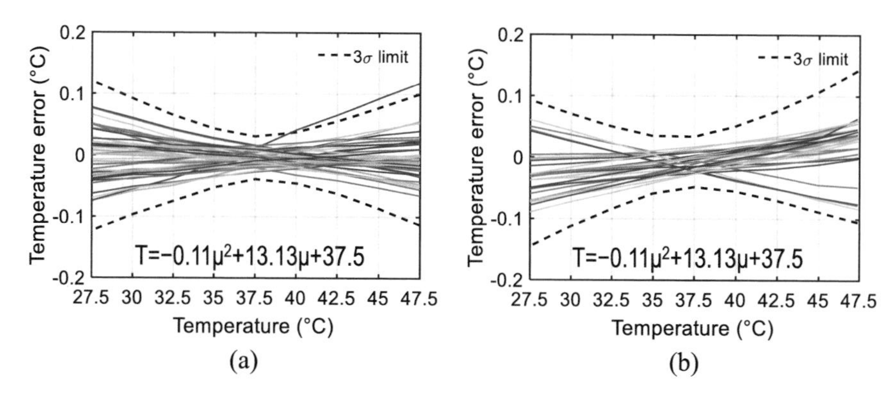

Fig. 5.8 Temperature inaccuracy with a quadratic master curve of (**a**) ceramic-packaged sensors (**b**) plastic-packaged sensors

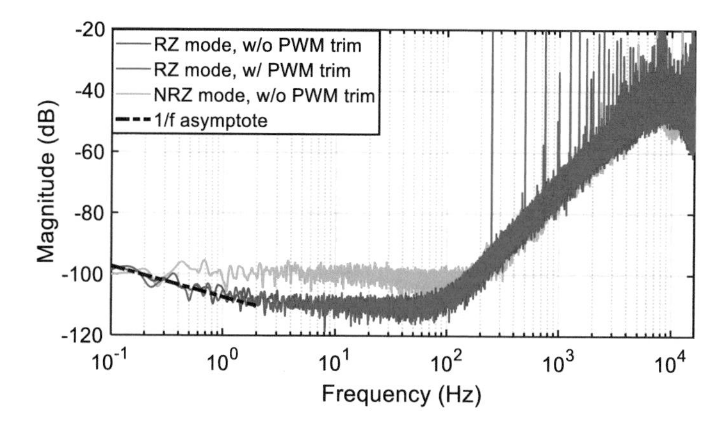

Fig. 5.9 Measured bitstream (BS) spectra with different settings

Even better accuracy can be achieved by using a quadratic master curve at the expense of a more complex digital backend. This results in a residual spread of below ±0.15 °C (3σ) for both ceramic- and plastic-packaged sensors, as shown in Fig. 5.8.

5.2.3.2 Resolution and FoM

FFTs of the sensor's bitstream (BS) outputs are shown in Fig. 5.9. As expected, the use of an NRZ DAC instead of a RZ DAC significantly degrades the modulator's noise floor (by 9 dB). Although PWM trimming does not impact the sensor's noise floor, it does introduce strong high-frequency tones. By limiting the conversion time (T_{conv}) to multiples of 8 ms, these tones can be completely filtered out by the notches

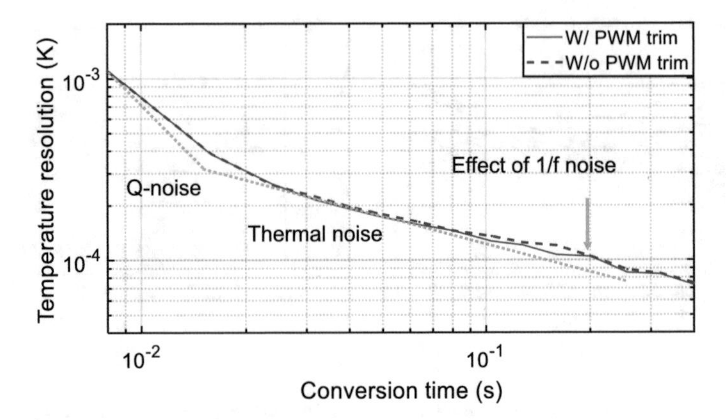

Fig. 5.10 Measured bitstream (BS) spectra with different settings

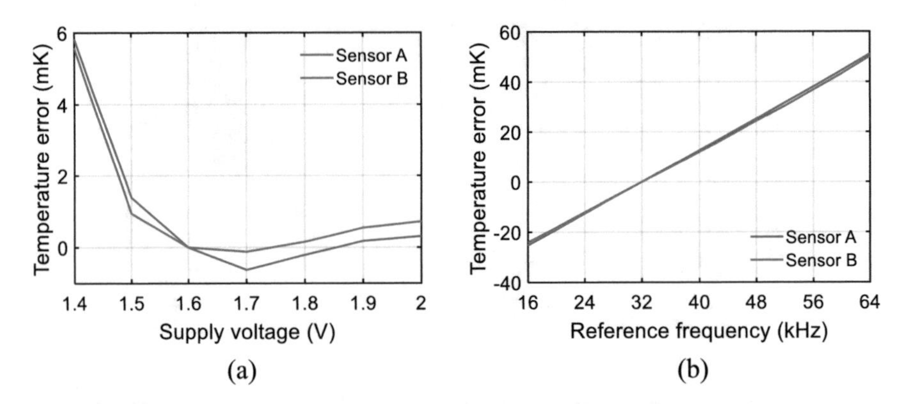

Fig. 5.11 (**a**) Supply sensitivity and (**b**) clock frequency sensitivity of two sensors on the same chip

of the sinc^2 decimation filter. For T_{conv} = 8 ms/40 ms, the sensor achieves 1.1 mK/200 μK (rms) resolution (Fig. 5.10), corresponding to resolution FoMs of 65 fJ·K^2/11 fJ·K^2.

5.2.3.3 Supply and Clock Sensitivity

As shown in Fig. 5.11a, the sensor achieves a power-supply sensitivity of only 4 mK/V from 1.5 V to 2 V at 37.5 °C (box method), which is the lowest ever reported for a temperature sensor. Its output is also robust to both input clock inaccuracy (1.6 mK/kHz) and clock jitter (<10% worse resolution with 2.2 ns cycle-to-cycle jitter), as shown in Fig. 5.11b. This makes the sensor well suited for use in wearable/implantable devices which often lack stable power supplies or well-defined clocks.

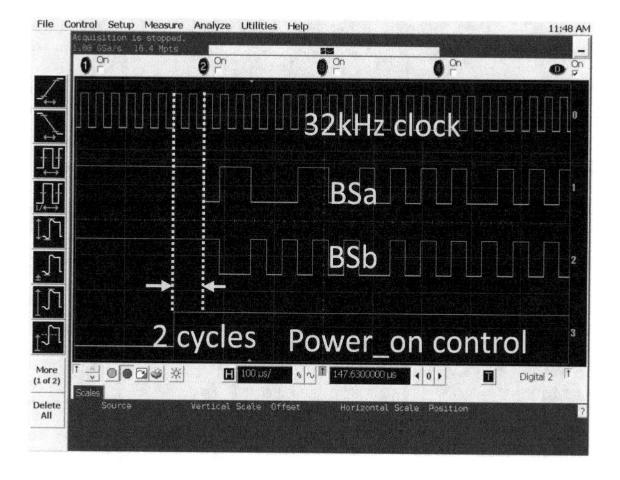

Fig. 5.12 Measured BS outputs of a chip around the rising edge of power-on control

5.2.3.4 Power-Down Mode

In power-down mode, the sensor only draws 125 nA at 37.5 °C. As shown in Fig. 5.12, the sensor starts up within 2 clock cycles (62.5 μs) of the rising edge of the power-on signal, thus facilitating efficient duty-cycling. At 10 conversions/s and T_{conv} = 8 ms, it dissipates an average power of only ~700 nW.

5.2.3.5 Comparison to Previous Work

Table 5.1 summarizes the performance of the proposed temperature sensor and compares it with the prior art. Compared to previous low-power designs [2–6], it achieves much higher resolution and a state-of-the-art resolution FoM. Moreover, the use of on-chip trimming means that, apart from a decimation filter, no further digital hardware, for example, for polynomial linearization or correlated trimming, is required. In contrast, after an offset trim, the measured inaccuracy of the sensor presented in Chap. 4, Sect. 4.6 is limited by nonlinearity and sensitivity variations and is about −0.6 °C/+0.2 °C (3σ) over the same temperature range.

5.2.4 Summary

A compact low-power resistor-based temperature sensor for biomedical purposes has been implemented in a standard 0.18 μm technology. It is built around a high resistance (600 kΩ) Wheatstone bridge that is read out in a self-balanced manner by a continuous-time delta-sigma modulator. An appropriately designed 1-bit series DAC improves both sensor nonlinearity and chip area, while a PWM-based trim reduces sensor spread and greatly simplifies its digital backend. The sensor

Table 5.1 Performance summary of the proposed sensor and comparison with the prior art

	[2]	[3]	[4]	[5]	[6]	**This work**	
Sensor type	BJT	MOS	DTMOS	Resistor	Resistor	**Resistor**	
CMOS Technology [nm]	180	350	160	180	65	**180**	
Area [mm²]	0.198	0.084	0.085	0.09	0.084	**0.12**	
Temperature range	25 °C to 45 °C	35 °C to 45 °C	−40 °C to 125 °C	−0 °C to 100 °C	−10 °C to 120 °C	**27.5 °C to 47.5 °C**	
3σ inaccuracy [°C] (trimming points)	±0.2 (1)	±0.1[a] (2)	±0.4 (1)	+1.5/−1.4[a] (2)	+0.34/−0.29[a] (2)	**+0.2/−0.1 (1)**	
Number of samples	20	3	16	18	10	**42**	
On-chip trim	No	No	No	No	No	**Yes**	
Supply voltage (V)	1.0 & 1.8	1.4 & 2.1	0.85	1.2	0.6 & 1	**1.6**	
Supply sensitivity (°C/V)	N.A.	0.3	0.45	14	1.0	**0.004**	
Power consumption [μW]	1.1	0.11	0.6	0.071	0.000346	**6.6**	
Conversion time [ms]	500	100	6	30	1000	**8**	**40**
Resolution [mK]	10	35	63	300	380	**1.1**	**0.20**
Resolution FoM [pJ·K²]	55	13	14.1	190	50	**0.064**	**0.011**

[a]Min or max

occupies 0.12 mm² and consumes only 6.6 μW from a 1.6 V supply. Additionally, it achieves a resolution FoM of 11 fJ·K² and an inaccuracy of +0.2 °C/−0.1 °C (3σ) in a ±10 °C range around body temperature. These results demonstrate that the proposed sensor can serve as an energy-efficient replacement for BJT- or MOS- based temperature sensors in biomedical applications.

5.3 A Wheatstone Bridge Sensor Embedded in a RC Frequency Reference[2]

5.3.1 Background Introduction

In order to reduce the volume and cost of generating a system clock, MEMS/BAW-based frequency references are widely used. An even better alternative would be the use of fully integrated CMOS-compatible frequency references. Depending on the

[2]H. Jiang, S. Pan, Ç. Gürleyük and K. A. A. Makinwa, "A 0.14mm² 16MHz CMOS RC Frequency Reference with a 1-point Trimmed Inaccuracy of ±400 ppm from −45 °C to 85 °C," in *IEEE*

time constant choices, these can be roughly categorized into three groups: LC oscillators [8, 9], thermal-diffusivity-based oscillators [10, 11], and RC oscillators [12, 13].

LC oscillators, since they are based on lithographically defined components, can achieve sub-100-ppm inaccuracy over the industrial temperature range from −40 °C to 85 °C [8, 9]. However, due to the frequency-dependent Q-factor of their on-chip inductors, such oscillators typically operate at GHz frequencies, and thus require mWs of power. Thermal-diffusivity-based oscillators can achieve an inaccuracy of ~300 ppm after proper compensation [10] and can operate at a much lower frequency (~10 MHz). However, they require on-chip heaters to generate thermal signals, which also consume several mWs.

RC-based oscillators dissipate much less power. Due to the large TCs of on-chip resistors (usually >100 ppm/°C), well-designed compensation is required to suppress the frequency error. In conventional designs, a rough 1st-order TC compensation can be achieved by combining resistors of positive/negative TCs in the analog domain [12, 13]. However, the high-order TCs will remain, which limits the achievable frequency inaccuracy.

Alternatively, the compensation can be achieved by digital temperature compensation after digitizing the RC time constant. Thus, an on-chip temperature sensor is required to provide a compensating polynomial. The minimum frequency error is then partially limited by the temperature sensor's inaccuracy.

This subsection describes a Wheatstone bridge temperature sensor embedded in such an RC frequency reference. To save chip area and power, a single ADC is multiplexed between the readout of RC filter and that of a Wheatstone sensor. Furthermore, the nonsilicided poly resistors in the RC filter and the Wheatstone bridge are shared, resulting in better accuracy since the spread of these resistors will then be correlated and so will partially cancel each other out. The frequency reference is realized in a standard 0.18 μm technology and occupies only 0.142 mm^2. After a single-point trim, the 16 MHz frequency reference achieves an inaccuracy of ±400 ppm from −45 °C to 85 °C.

5.3.2 Circuit Implementation

5.3.2.1 Circuit Principle

The basic principle of the digitally compensated RC frequency reference is shown in Fig. 5.13 [14]. It consists of a frequency-locked loop (FLL), which locks the frequency f_{DCO} of a digitally controlled oscillator (DCO) to the phase shift of an RC filter. The temperature dependence of the RC filter is digitally compensated by the information provided by a temperature sensor. To relax its accuracy requirements, the temperature coefficient (TC) of the RC filter should be minimized. In this work,

---

ISSCC Dig. Tech. Papers., Feb. 2021, pp. 436-437.

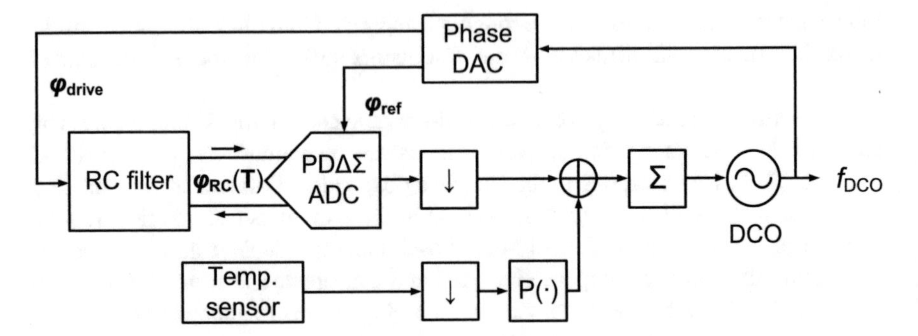

Fig. 5.13 Block diagram of a digitally compensated RC frequency reference

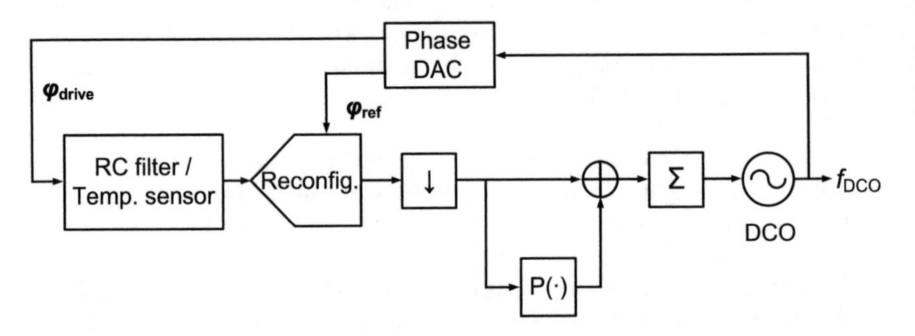

Fig. 5.14 Block diagram of the proposed RC frequency reference

it is implemented with nonsilicided p-poly resistors (TC ~ −240 ppm/°C) and MIM capacitors (TC ~ −30 ppm/°C).

To minimize the impact of resistor/capacitor spread on frequency error, as well as to save circuit area, the temperature sensor is merged with the RC filter. Furthermore, a single reconfigurable ΔΣ-ADC is reused to readout both of them, as shown in Fig. 5.14.

5.3.2.2 Reconfigurable RC Network and ADC

Figure 5.15 shows the simplified schematic diagram of the reconfigurable RC network and the ADC. It employs a WhB temperature sensor made up of p-poly resistors (R_n, 128 kΩ, TC ~ −240 ppm/°C) and silicided-diffusion resistors (R_p, 92 kΩ, TC ~ 3000 ppm/°C). To reduce area and minimize the number of components that contribute to spread, the RC filter is realized by combining a single MIM capacitor (C_0) with the R_n branches of the WhB. The result is a low-pass filter (LPF) whose phase shift is determined by a single RC time constant. Compared to WBs, whose

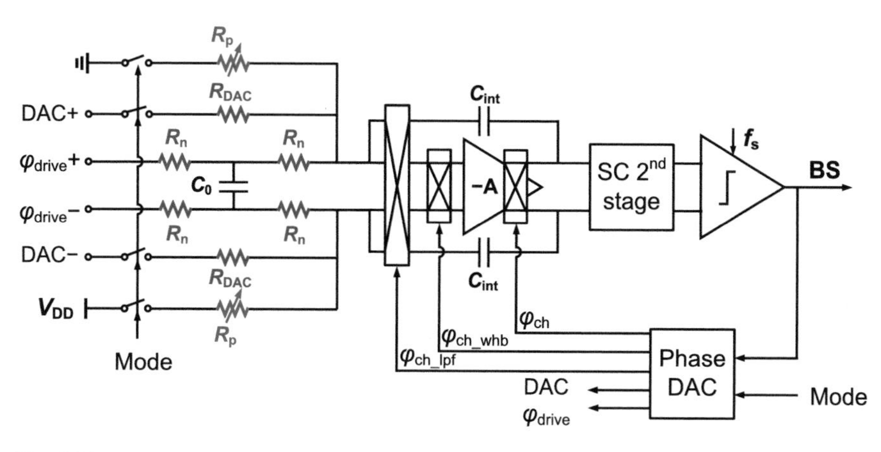

Fig. 5.15 Simplified circuits diagram of the LPF/WhB readout circuit

phase response is determined by the interaction of two different time constants, the use of a LPF is a promising way to achieve better inaccuracy after a 1-point trim. As in [14], the output of the WhB sensor and the phase shift of the LPF exhibit similar nonlinearity over the target temperature range (−45 °C to 85 °C), so that good accuracy can be achieved with a 4th-order nonlinearity-correction polynomial. Furthermore, due to the shared p-poly resistors, the effect of their TC spread on the LPF and the WhB is correlated, resulting in a significant reduction in the spread of the temperature-compensated phase shift of the LPF.

Both the LPF and the WhB are digitized by a feedforward 2nd-order CTΔΣ-ADC, with the 1st-stage amplifier chopped at its sampling frequency (f_s). In the LPF mode, both R_p and R_{DAC} are disconnected from the supply, and the LPF is driven by a square wave at $f_{drive} = f_s = 500$ kHz, which, after calibration, will be eventually provided by the DCO. By enabling the LPF chopper at the input of the 1st-stage amplifier (φ_{ch_lpf}) while disabling that of the WhB (φ_{ch_whb}), the CTΔΣ-ADC is configured to demodulate the phase output of the LPF (Fig. 5.16a), like that presented in Chap. 3. To maximize phase sensitivity of the LPF, the phase references for demodulation (φ_{ch} and φ_{ch_lpf}) is set to −45° ± 5.625° w.r.t φ_{drive}, that is, nearly orthogonal to the LPF's center frequency.

In the WhB mode (Fig. 5.16b), φ_{ch_whb} and φ_{drive} are disabled. The BS output toggles the RDAC switches the same way as introduced Chap. 4. To avoid the extra supply current coming from DAC resistors, the DAC resistors are switched between supply rails and V_{CM} (Chap. 4, Sect. 4.6). Due to the DC operation in the WhB mode, the LPF capacitor (C_0) will not contribute any error.

All the circuit blocks of the reconfigurable ΔΣ-ADC, including the 1st-stage amplifier and the switched-capacitor 2nd stage, are reused from the 2nd WB sensor implementation (Chap. 3, Sect. 3.4). The current of the 1st-stage amplifier (~8 μA) is scaled to efficiently process the maximum current of the resistive front-end. The design of other blocks (polynomial engine and DCO) is beyond the scope of this book.

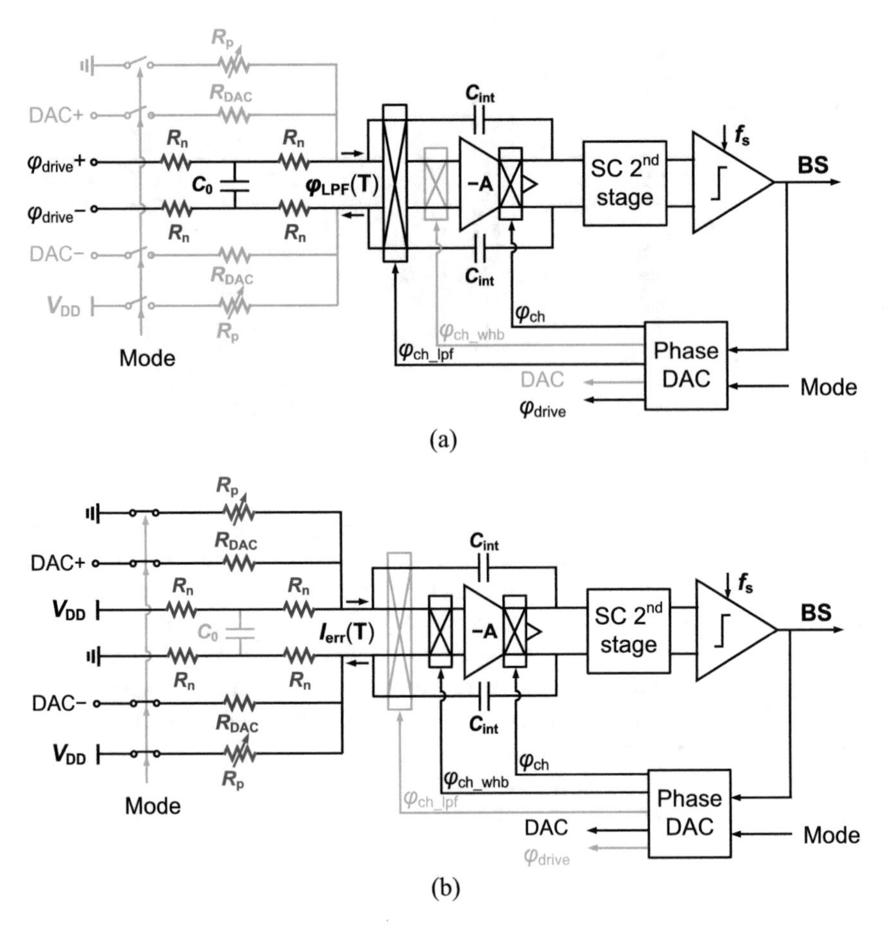

Fig. 5.16 Simplified schematic diagram under (**a**) the LPF mode and (**b**) the WhB mode

5.3.3 Measurement Results

The prototype was fabricated in a standard 0.18 μm CMOS process and packaged in ceramic DIL (Fig. 5.17). It occupies 0.142 mm^2, including ΔΣ-ADC with WhB & LPF, and DCO, while the former takes 88% (0.125 mm^2). However. as the silicided diffusion resistors in the WhB sensor operate at DC, it can be comfortably placed below the 1st-stage integrator capacitor of the ΔΣ-ADC and avoid occupying extra area. Also, the p-poly resistors in the WhB are reused by the LPF. As a result, the additional area required by the WhB sensor is negligible. The circuit draws 88 μA from a 1.8 V voltage supply in the normal operation mode, or 41 μA in the WhB readout mode. For flexibility, the digital backend was implemented in an external FPGA.

Fig. 5.17 Die photo of the fabricated chip

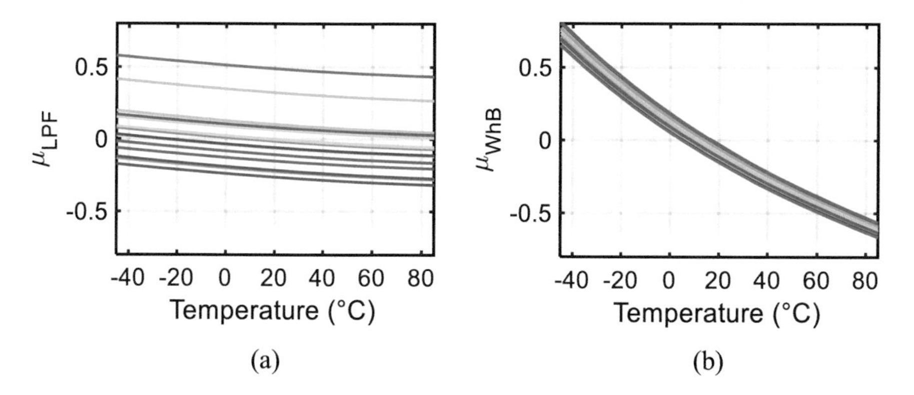

Fig. 5.18 Measured bitstream averages of (**a**) LPFs and (**b**) WhB sensors using an external 16 MHz frequency reference

5.3.3.1 Calibration and Inaccuracy

Fifteen devices were characterized in a temperature-controlled oven from −45 °C to 85 °C (Fig. 5.18). Without trimming, the WhB sensor's peak-to-peak inaccuracy is ~14 °C. After an offset trim and a fixed 4th-order polynomial correction, the WhB

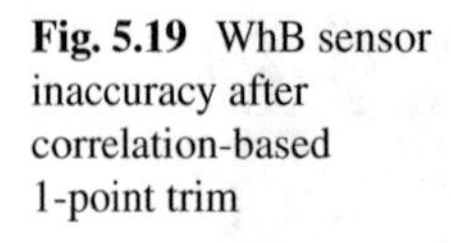

Fig. 5.19 WhB sensor inaccuracy after correlation-based 1-point trim

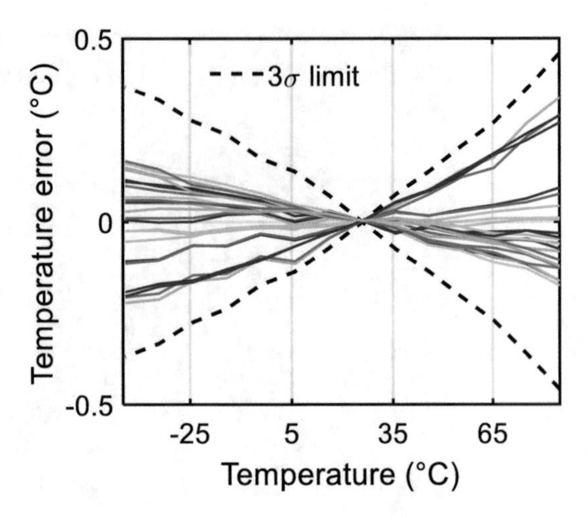

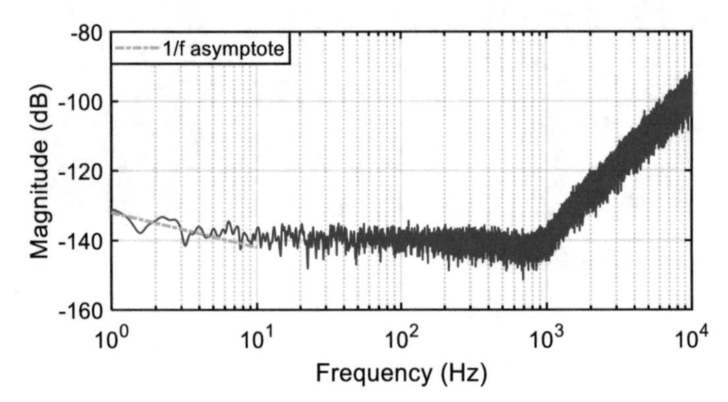

Fig. 5.20 Power spectral density of the bitstream output

sensor achieves a 3σ inaccuracy of 4.2 °C. This can be improved to 0.46 °C with a correlated 1-point calibration, as shown in Fig. 5.19.

5.3.3.2 Resolution and FoM

The bitstream spectra of the WhB sensor is shown in Fig. 5.20. The observed $1/f$ corner is ~10 Hz, which is mainly from the nonsilicided poly resistors. Due to the lack of an identical sensor on the same chip, the sensor's resolution is determined by calculating the standard deviation of data from a single sensor after drift compensation. Within a 1 s interval, the calculated resolution using sinc^2 filters is shown in Fig. 5.21. With a conversion time of 10 ms, its 200 μK_{rms} resolution corresponds to a resolution FoM of 30 fJ·K^2.

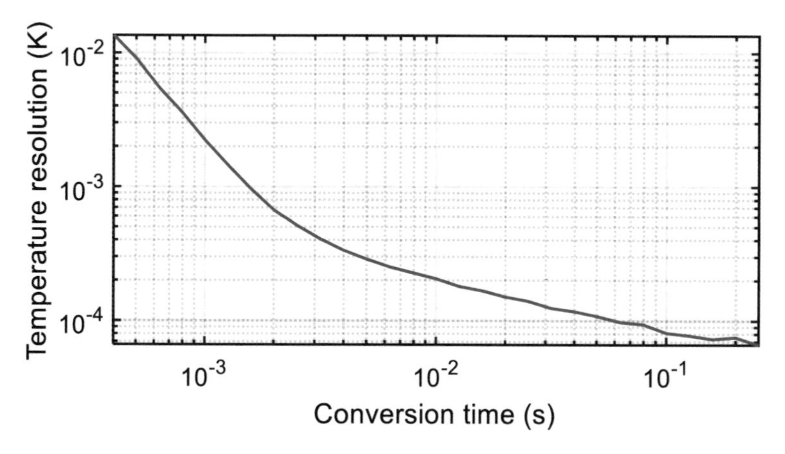

Fig. 5.21 Temperature resolution versus conversion time

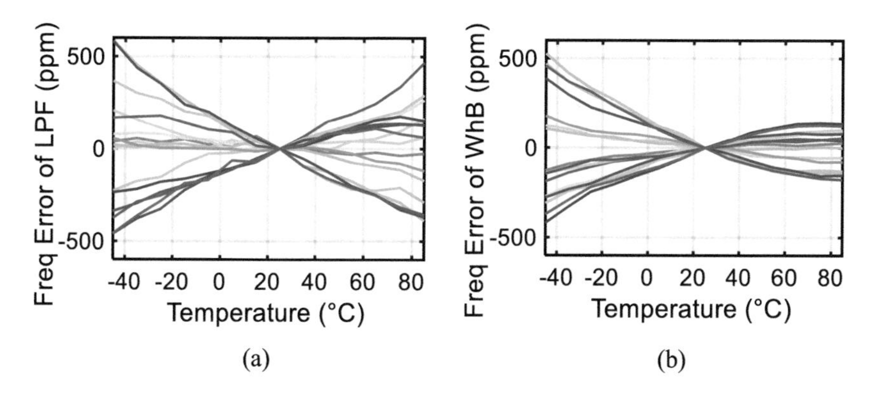

Fig. 5.22 Estimated residue frequency error due to (**a**) LPF and (**b**) WhB

5.3.3.3 Frequency Reference

After closing the loop of the frequency reference chip, the samples were characterized from −45 °C to 85 °C. As expected, the temperature dependence of the LPF output is quite low (−240 ppm/°C). Compared to WB sensors presented in Chap. 3, the reduced ±5.625° phase DAC range introduces much smaller nonlinearity. From simulations, it contributes to less than 100 ppm error after a simple offset trim, and the complicated μ to RC mapping can be skipped. Since the LPF nonlinearity is similar to that of the WhB output (Fig. 5.18), most of the frequency error can be removed by applying a 1-point offset trim to the outputs of both the WhB and LPF at room temperature. As the WhB and the LPF are implemented on the same die, this trim is quite robust to ambient temperature variations [14]. After trimming, open-loop measurements show that the estimated frequency error due to the LPF alone is about ±520 ppm, while the error due to the WhB alone is about ±470 ppm, as shown in Fig. 5.22. Due to the shared p-poly resistors, these errors are somewhat

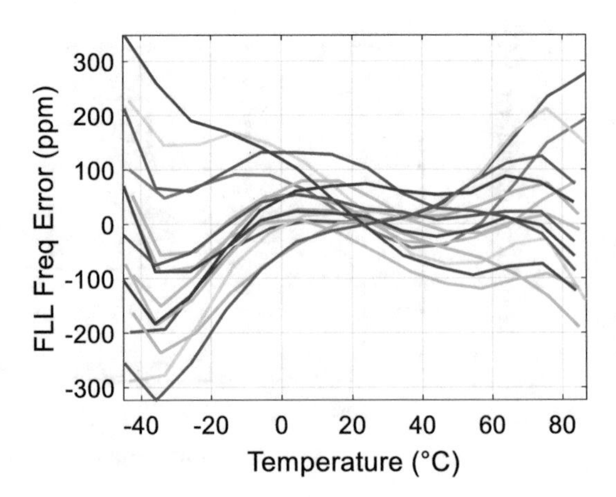

Fig. 5.23 Measured frequency error after a 1-p trim and 4th-order systematic nonlinearity correction

Table 5.2 Performance summary of the WhB sensor and comparison with those presented in Chap. 4

	JSSC'15 [15]	Implementation III	Implementation IV		**This work**
Sensor type	Resistor WhB	Resistor WhB	Resistor WhB		**Resistor WhB**
Technology	0.18 μm	0.18 μm	0.18 μm		**0.18 μm**
Area [mm²]	0.43	0.12	0.11		**0.125[a]**
Temp. Range [°C]	−40—125	−55—125	−55—125		**−45—85**
3σ Inaccuracy [°C] (trimming points)	0.4[b] (2[c])	0.14 (2[d])	0.1 (2[d])	0.4 (1)	**0.46 (1)**
Relative inaccuracy	0.48%	0.16%	0.11%	0.44%	**0.7%**
Power [μW]	65	79	55		**74**
Conv. time [ms]	0.1	10	8		**10**
Resolution [mK]	10	0.16	0.15		**0.20**
Res. FoM[fJ·K²]	650	20	10		**30**

[a]Mostly reused area
[b]Min/Max
[c]1-point trim + 1st-order fit
[d]1st-order fit

correlated, resulting in lower frequency error when the loop is closed. After an individual 1-point trim and a fixed 4th-order polynomial nonlinearity correction, the proposed frequency reference achieves better than ±400 ppm inaccuracy (Fig. 5.23), which corresponds to a residual TC of 6.2 ppm/°C (box method). The closed-loop period jitter is 10.2ps_{rms}. With a fixed digital input, the Allan Deviation of the DCO alone is around 40 ppm, which drops to 0.4 ppm in the closed-loop configuration.

5.3.3.4 Comparison to Previous Work

Table 5.2 summarizes the performance of the proposed WhB sensor and compares it with sensors presented in Chap. 4. Due to the use of a single-bit DAC, the ΔΣ-ADC's 1st stage is not optimized for noise. Also, the WhB sensitivity is smaller due to the use of a low-TC poly resistor. Consequently, the sensor's energy efficiency is worse than that of previous WhB sensors. Also, its 1-point trim accuracy is slightly worse. However, this sensor is well suited for the RC frequency reference compensation application, as it requires almost no additional chip area, while the accuracy/resolution performance is good enough.

The performance summary of the frequency reference is summarized in Table 5.3 and compared to other high-accuracy designs that achieve <10 ppm/°C residual TC. Compared to [14], the proposed frequency reference achieves comparable accuracy after a 1-point trim at room temperature, as well as better energy and area efficiency. This makes it highly compatible with IoT and wireline applications that require good accuracy and a high level of integration.

Table 5.3 Performance summary of the RC frequency reference and comparison with state of the art

	ISSCC'20 [14]		ISSCC'20 [16]	JSSC'18 [17]	JSSC'18 [18]	**This Work**
Technology	0.18 μm		65 nm	0.18 μm	0.18 μm	**0.18 μm**
Area [mm²]	0.3		0.18	1.65	0.17	**0.14**
Frequency [MHz]	16		32	7	24	**16**
Norm. Inaccuracy [ppm]	±400	±100	±530	±170	±215	**±400**
Trimming Points	2	2+Batch (6th order)	2	2+Batch (4th order)	3	**1+Batch (4th order)**
Temperature Range [°C]	−45 to 85		−45 to 85	−45 to 85	−40 to 150	**−45 to 85**
Supply Range [V]	1.6 to 2.0		1.1 to 3.3	1.7 to 2.0	1.8 to 5.0	**1.6 to 2.0**
Supply Sensitivity [%/V]	0.12		0.008[a]	0.18	0.01	**0.2**
Number of Samples	20		6	8	1	**12**
Allan Deviation [ppm]	0.32		2.5	0.33	–	**0.4**
Power [μW]	400		34	750	200	**158.4**

[a]with on-chip LDO

5.3.4 Summary

A Wheatstone bridge resistor–based temperature sensor has been implemented in a standard 0.18 μm technology for the temperature drift compensation of an RC-based frequency reference. By reusing the existing readout circuit of the RC filter and doing careful layout, the sensor requires almost no additional chip area. After a 1-point trim, the frequency reference achieves an inaccuracy of better than ±400 ppm from −45 °C to 85 °C. Compared to the fourth implementation shown in Chap. 4, the degraded 1-point trim inaccuracy is possibly due to a worse fabrication quality: the no-trim inaccuracy of this sensor is about 2× worse, and so is the error after a 1-point trim.

5.4 Concluding Remarks

In this chapter, two resistor-based temperature sensor prototypes are presented to demonstrate their feasibility in different applications: biomedical temperature sensing and temperature compensation of RC frequency references. There are many other applications of resistor-based temperature sensors, such as thermal monitoring of microprocessors/DRAMs and wireless sensor nodes. These can be investigated in future work.

References

1. *Standard Specification for Electronic Thermometer for Intermittent Determination of Patient Temperature*, Standard ASTM E1112-00 (2018)
2. M. Law, S. Lu, T. Wu, A. Bermak, P. Mak, R.P. Martins, A 1.1 μW CMOS smart temperature sensor with an inaccuracy of ±0.2 °C (3σ) for clinical temperature monitoring. IEEE Sensors J. **16**(8), 2272–2281 (2016)
3. A. Vaz, A. Ubarretxena, I. Zalbide, D. Pardo, H. Solar, A. García-Alonso, R. Berenguer, Full passive UHF tag with a temperature sensor suitable for human body temperature monitoring. IEEE TCAS II **57**(2), 95–99 (2010)
4. K. Souri, Y. Chae, F. Thus, K. Makinwa, A 0.85V 600nW all-CMOS temperature sensor with an inaccuracy of ±0.4°C (3σ) from −40 to 125°C, in *IEEE ISSCC Dig. Tech. Papers*, (2014, Feb), pp. 222–223
5. S. Jeong, Z. Foo, Y. Lee, J. Sim, D. Blaauw, D. Sylvester, A fully-integrated 71 nW CMOS temperature sensor for low power wireless sensor nodes. IEEE J. Solid State Circuits **49**(8), 1682–1693 (2014)
6. H. Xin, M. Andraud, P. Baltus, E. Cantatore, P. Harpe, A 0.34-571nW all-dynamic versatile sensor interface for temperature, capacitance, and resistance sensing, in *Proc. IEEE ESSCIRC*, (2019, Sept), pp. 161–164
7. S. Hacine, T.E. Khach, F. Mailly, L. Latorre, P. Nouet, A micropower high-resolution ΣΔ CMOS temperature sensor, in *Proc. IEEE Sensors*, (2011, Oct), pp. 1530–1533
8. M.S. McCorquodale et al., A 25-MHz self-referenced solid-state frequency source suitable for XO-replacement. IEEE Trans. Circuits Syst. I, Reg. Papers **56**(5), 943–956 (2009)

9. E.O. Ates, A. Ergul, D.Y. Aksin, Fully integrated frequency reference with 1.7 ppm temperature accuracy within 0–80°C. IEEE J. Solid State Circuits **48**(11), 2850–2859 (2013)
10. L. Pedalà, Ç. Gürleyük, S. Pan, F. Sebastiano, K.A.A. Makinwa, A frequency-locked loop based on an oxide electrothermal filter in standard CMOS, in *IEEE Proc. ESSCIRC*, (2017), pp. 7–10
11. S.M. Kashmiri, K. Souri, K.A.A. Makinwa, A scaled thermal diffusivity- based 16 MHz frequency reference in 0.16 μm CMOS. IEEE J. Solid State Circuits **47**(7), 1535–1545 (2012)
12. T. Jang, M. Choi, S. Jeong, S. Bang, D. Sylvester, D. Blaauw, A 4.7 nW 13.8 ppm/ °C self-biased wakeup timer using a switched-resistor scheme, in *IEEE ISSCC Dig. Tech. Papers*, (2016, Feb), pp. 102–103
13. D. Griffith, P.T. Røine, J. Murdock, R. Smith, A 190 nW 33 kHz RC oscillator with ± 0.21% temperature stability and 4 ppm long-term stability, in *IEEE ISSCC Dig. Tech. Papers*, (2014, Feb), pp. 300–301
14. Ç. Gürleyük, S. Pan, K.A.A. Makinwa, A 16MHz CMOS RC frequency reference with ±400ppm inaccuracy from −45°C to 85°C after digital linear temperature compensation, in *IEEE ISSCC Dig. Tech. Papers*, (2020, Feb), pp. 64–66
15. C.H. Weng, C.K. Wu, T.H. Lin, A CMOS thermistor-embedded continuous-time delta-sigma temperature sensor with a resolution FoM of 0.65 pJ °C^2. IEEE J. Solid State Circuits **50**(11), 2491–2500 (2015)
16. A. Khashaba et al., A 34μW 32MHz RC oscillator with ±530ppm inaccuracy from −40°C to 85°C and 80ppm/V supply sensitivity enabled by pulse-density modulated resistors, in *IEEE ISSCC Dig. Tech. Papers*, (2020, Feb), pp. 66–68
17. Ç. Gürleyük et al., A CMOS dual-RC frequency reference with ±200-ppm inaccuracy from −45°C to 85 °C. IEEE J. Solid State Circuits **53**(12), 3386–3395 (2018)
18. G. Zhang, K. Yayama, A. Katsushima, T. Miki, A 3.2 ppm/°C second-order temperature compensated CMOS On-Chip oscillator using voltage ratio adjusting technique. IEEE J. Solid State Circuits **53**(4), 1184–1191 (2018)

Chapter 6
Conclusions and Outlook

In this book, the development of resistor-based temperature sensors for the compensation of MEMS frequency references and other applications has been described. The main findings are presented in this chapter. Furthermore, the performance of resistor-based temperature sensors is summarized and compared to that of other types of CMOS temperature sensors, and some future research directions are proposed.

6.1 Main Findings

The main findings of this work are as follows:

- Of the various types of resistors available in CMOS processes, the silicided resistor is the best suited for temperature sensing due to its large temperature coefficient (TC) and good stability (Chap. 1).
- Due to the relatively large TC of on-chip resistors, the theoretical energy efficiency of resistor-based temperature sensors is over an order of magnitude larger than that of conventional BJT-based ones (Chap. 1).
- Depending on the choice of the impedance reference, resistor-based temperature sensors have different characteristics. RC-based sensors, due to the use of temperature-insensitive capacitors, can achieve high accuracy, while dual-R-based sensors can achieve high energy efficiency, especially by employing resistors with TCs of opposite polarity (Chap. 2).
- Wien bridge (WB) and Wheatstone bridge (WhB) sensors are the preferred implementations of RC- and dual-R-based sensors, respectively (Chap. 2).

S. Pan, K. A. A. Makinwa, *Resistor-based Temperature Sensors in CMOS Technology*, ACSP · Analog Circuits and Signal Processing,
https://doi.org/10.1007/978-3-030-95284-6_6

- Despite exhibiting large untrimmed errors (>10 °C), Wien bridge sensors can achieve an inaccuracy of around 0.03 °C over a 180 °C temperature range with only two trimming points (Chap. 3).
- Compared to other types of poly resistors, silicided poly resistors have a much lower 1/*f* noise level. Also, they have lower stress dependency: after a 2-point trim, the error introduced from plastic packaging is less than 0.2 °C over a 220 °C temperature range (Chap. 3).
- State-of-the-art energy efficiency can be obtained by optimizing both the front-end and the readout circuit of a Wheatstone bridge sensor. A 10 fJ·K^2 FoM has been achieved, which is only 6× larger than the theoretical limit imposed by the thermal noise of the Wheatstone bridge (Chap. 4).
- Although Wheatstone bridge sensors have a large nonlinearity (>15 °C over a 180 °C temperature range) and thus require high-order polynomial correction, this nonlinearity is quite stable: varying by <0.1 °C over 3 batches (Chap. 4).
- By simply inserting tail resistors in a pseudo-differential OTA, the nonlinearity of its transconductance can be reduced by >10× over corners without degrading its noise performance (Chap. 4).
- The use of a tail-resistor linearized OTA greatly improves the efficiency of multi-bit CTΔΣ-ADCs. The improved linearity leads to a significant reduction in quantization-noise folding, allowing the tail current of the OTA to be optimized for thermal noise (Chap. 4).
- Apart from the frequency compensation of MEMS/crystal oscillators, resistor-based temperature sensors can be advantageously used in other applications, for example, in biomedical systems and RC-based frequency references (Chap. 5).

6.2 Temperature Sensor Comparison

To demonstrate the place of the presented resistor-based sensors in the temperature sensor universe, the best specifications of different types of sensors are summarized in Table 6.1. As the numbers were picked from different publications, design trade-offs (e.g., area vs. no-trim inaccuracy) are not reflected in this table.

As shown in Table 6.1, the WB/WhB sensors presented in this book can achieve state-of-the-art performance in terms of temperature range, relative inaccuracy, supply sensitivity, resolution, and resolution FoM. The designs were not particularly optimized for area and supply voltage, so this can still be improved. Compared to all CMOS temperature sensors published to date (2020), the Wien bridge sensor presented in Sect. 3.5 achieves the best inaccuracy after a 2-point trim, the Wheatstone bridge sensor shown in Sect. 4.6 achieves the best resolution FoM, while the Wheatstone bridge sensor described in Sect. 5.2 achieves the lowest supply sensitivity.

Table 6.1 State-of-the-art specifications of different types of CMOS temperature sensors circa 2020 [1]

	BJT	MOS	RES			TD	Best reported
			WB	WhB	Overall		
Area [mm^2]	0.0025	**0.001**	0.007	0.044	0.006	0.0017	MOS [4]
Temp. Range [°C]	255	200	220	180	220	**270**	TD [5]
Relative inaccuracy (no trim)	0.3%	-	-	-	-	**0.2%**	TD [6]
Relative inaccuracy (1-pt trim)	**0.06%**	0.44%	0.32%	0.44%	0.32%	-	BJT [7]
Relative inaccuracy (2-pt trim)	-	0.11%	0.03%	0.11%	**0.03%**	-	RES (WB, Chap.3)
Supply voltage [V]	0.5 [a]	**0.4**	1.8	0.7	0.6	1	MOS [8]
Supply sensitivity [°C/V]	0.008	0.27	0.17	0.004	**0.004**	-	RES (WhB, Chap.5)
Power [μW]	0.0005	0.00013	31	0.00005	**0.00005**	1300	RES [9]
Conversion time [ms]	0.01	**0.002**	5	0.01	0.0025	1	MOS [10]
Resolution [mK]	0.65	10	0.45	0.15	**0.1**	15	RES [11]
Res. FoM [pJ·K^2]	0.19	0.26	0.11	0.01	**0.01**	140000	RES (WhB, Chap.4)

[a] With a charge-pump.

The shaded blocks indicate results presented in this work, while the numbers in bold are the best reported by any type of CMOS sensor

The energy efficiency of various types of CMOS temperature sensors is compared in Fig. 6.1. Compared to Fig. 1.2, solid star/square symbols have been added to indicate sensors presented in this work, as well as those published by others between 2017 and 2020. Prior to the start of this research, the best-reported resolution FoM for a CMOS temperature sensor was 650 fJ·K^2 [2], as denoted by the dashed line. By systematic circuit optimization, this has been improved by 65×, as represented by the solid line (10 fJ·K^2). This FoM is even 4× better than that of MEMS-based sensors [3], which are not CMOS-compatible.

Roughly half of the specification records listed in Table 6.1 still belong to other types of sensors. Of these, inaccuracy is limited by the inherent properties of resistors, and will be hard to improve. Of the other specifications, however, there are no fundamental limitations to the further improvement of resistor-based sensors. For example, being a passive element, the maximum operating temperature range of resistors can be expected to be larger than that of BJTs or MOSFETs, while the minimum supply voltage should be smaller.

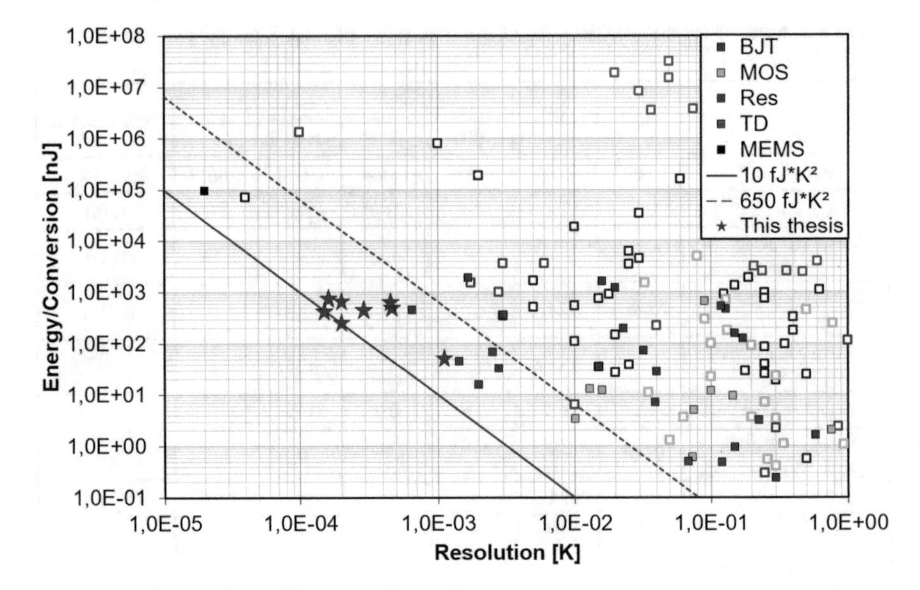

Fig. 6.1 Energy per conversion versus resolution of temperature sensors [1]. The stars represent the sensors presented in this book, while the solid/hollow squares indicate other sensors realized during/before this research, which started in 2016

6.3 Systematic Design Approaches for Accuracy

This book provides detailed design guidelines for improving the energy efficiency of resistor-based sensors. However, a similar approach to improving the accuracy of such sensors would also be useful. Apart from facilitating the design of sensors with better accuracy, this knowledge would also help when designing compact sensors with relaxed inaccuracy requirements.

6.3.1 Cadence Modeling

For BJT-based temperature sensors, one major source of inaccuracy is the effect of lithographic tolerances on the BJT's base width and the emitter area. This can be suppressed by using larger sensing BJTs [12]. Due to the existence of accurate process and mismatch models, the inaccuracy of BJT-based sensors can be well-predicted via simulations.

Unlike BJTs, the temperature characteristics of on-chip resistors are not as well-modeled. All the sensors presented in this book were implemented in the same standard 0.18 μm CMOS technology. According to its documentation, all resistor variations (process and mismatch) are modeled as changes in the sheet resistance. In practice, however, both the sheet resistance and TC will spread, and there could be even some correlation between them [13].

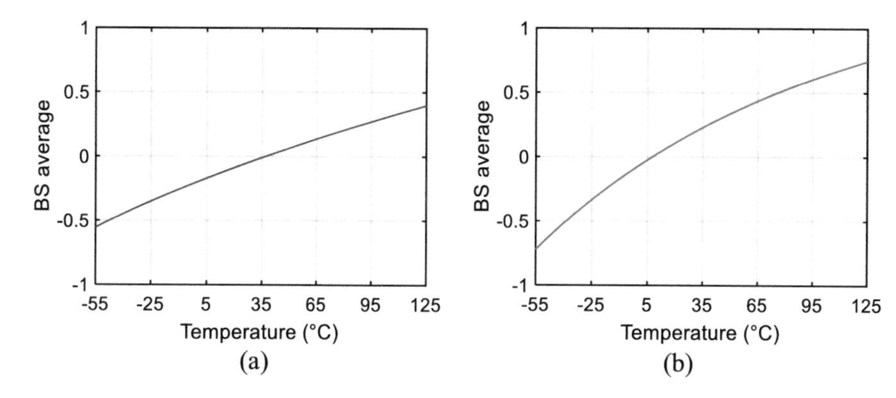

Fig. 6.2 Simulated output of (**a**) WB sensor and (**b**) WhB sensor with ideal readout electronics

To test the effectiveness of the process models, Monte Carlo simulations (device mismatch) have been done on the Wien bridge sensor presented in Sect. 3.5 and the Wheatstone bridge sensor presented in Sect. 4.6, as shown in Fig. 6.2. For simplicity, the readout circuits are assumed to be ideal. The untrimmed inaccuracies of both sensors are less than 1 °C, which is over 10× smaller than the measurement results. With process spread simulations, the untrimmed temperature errors become larger than 100 °C. However, as these simulations assume perfectly matched resistors and capacitors, the residual error will become zero after a single-point trim.

After applying the same 1-point/1st-order trimming methods used in Chaps. 3 and 4 on Fig. 6.2, the simulated sensor inaccuracies are on average an order of magnitude smaller than the measurement results, as shown in Figs. 6.3 and 6.4. The almost-zero residual error shown in Fig. 6.4b also confirms the theory presented in Chap. 4, Sect. 4.2.2, that, without TC variations of resistors, a Wheatstone bridge sensor can be perfectly calibrated with two trimming points.

There are two possible reasons for the large discrepancy between simulation and measurement. The first reason is that the readout circuits presented in this book may be significant sources of inaccuracy. The second, and more likely, reason is that the simplified resistor model results in over-optimistic simulation results.

6.3.2 Data Analysis

Another way to predict the achievable sensor inaccuracy is by analyzing all reported results. Figure 6.5 shows the inaccuracy versus resistor area of different sensors. To enlarge the data set, WB/WhB sensors are regarded as special cases of RC/dual-R sensors, respectively. In general, sensors that use silicide resistors are more accurate than those that use other resistors.

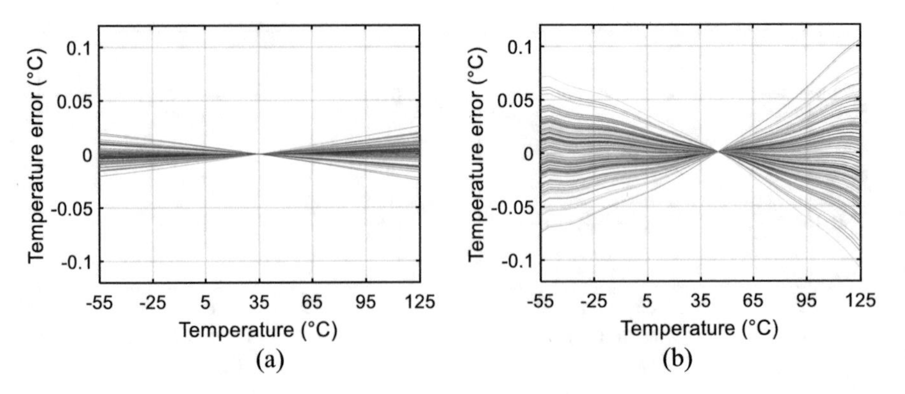

Fig. 6.3 Simulated inaccuracy of (**a**) WB sensor and (**b**) WhB sensor after a 1-point trim and systematic error removal

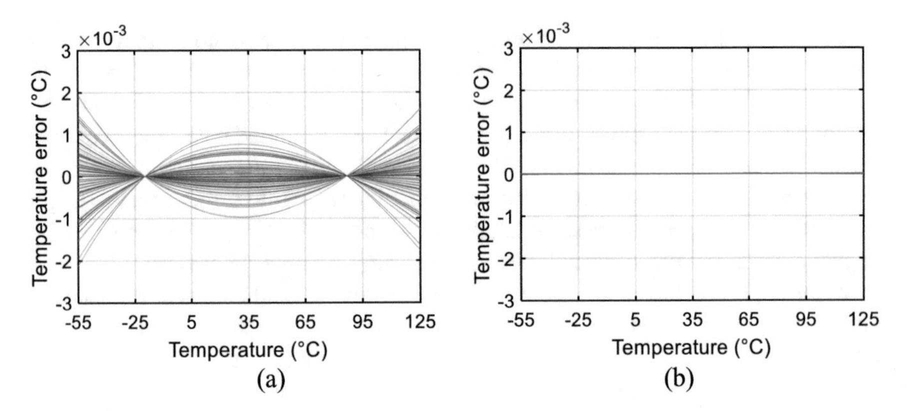

Fig. 6.4 Simulated inaccuracy of (**a**) WB sensor and (**b**) WhB sensor after 1st-order fit and systematic error removal

As can be seen from the rough envelope curves in Fig. 6.5a, the inaccuracy of RC sensors improves as the resistor area increases [14]. However, the trend stops when the resistor area reaches ~0.1 mm^2. This can be explained as a combined effect of inherent inaccuracy of resistors (nominal value and TC spread) and the device mismatch. The former dominates the inaccuracy with a large resistor area, while the latter is the major inaccuracy contributor for compact sensors.

As of 2020, very small silicide resistors have not been used in dual-R sensors. As a result, envelope curves in the inaccuracy versus area plot (Fig. 6.5b) cannot be drawn. It has been shown in [24] that small Wheatstone bridge sensors made from nonsilicided resistors can also achieve decent inaccuracy after a 2-point trim. With carefully designed readout circuits, similar sensors with silicide resistors should be even more accurate.

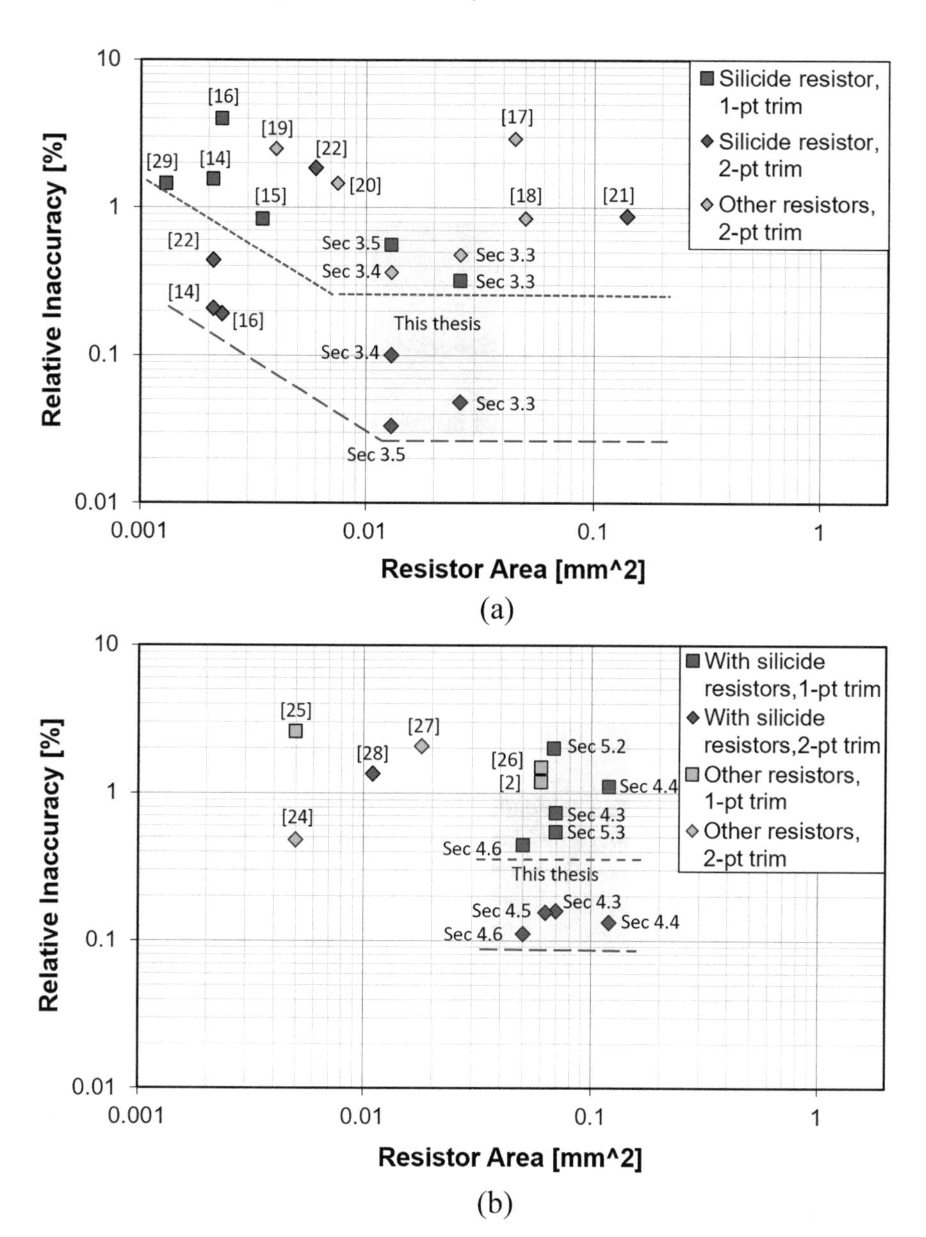

Fig. 6.5 Relative inaccuracy versus resistor area of (**a**) RC and (**b**) dual-R sensors. The resistor areas were estimated from the published chip photos

6.3.3 Experimental Verification

The direct way to investigate the achievable inaccuracy of resistor-based sensors is to build an array of different types of stand-alone resistors and characterize them carefully. Although a bit tedious, this approach will rule out the effect of imperfect readout circuits.

6.4 More Future Research Directions

Apart from the aforementioned issue of the relationship between resistor area and sensor inaccuracy, many other directions related to this book are also worth investigating. Here is an incomplete list.

6.4.1 Area- and Power-Efficient Digital Backend

For simplicity, the digital backends of all the sensors presented in this book are implemented off-chip. However, commercial chips should have all the digital integrated, so that they can use standard protocols, like SPI or I^2C, for data communication.

Designing such a digital back-end is a nontrivial task especially under the constraint that its area and power should not dominate that of the analog front-end. Take a Wien bridge sensor, for example, two high-order polynomials and a 2-point trim are required to remove its process spread and nonlinearity, and all the steps require high resolution. If not designed carefully, the digital calibration circuit alone would consume more area and power than the analog front-end.

6.4.2 Background Calibration of Wheatstone Bridge Sensors

As presented in Chap. 2, WhB sensors with unsilicided poly resistors are vulnerable to long-term drift. When used for the compensation of MEMS frequency references, this drift might reduce the long-term stability of such references to tens of ppm. To overcome this issue, the WhB sensor can be periodically calibrated by a more stable temperature sensor, for example, a BJT-based temperature sensor. As a result, the energy efficiency of the resulting hybrid sensor will still be determined by the WhB sensor, while the long-term stability will be limited by that of the BJT sensor, which can be less than 6 mK over one year [30]. To reduce chip area, and thus fabrication cost, the readout circuit should preferably be shared by the two sensors.

6.4.3 Long-Term Stability of Wien Bridge Sensors

Compared to WhB sensors, WB sensors are made from stable silicided poly resistors and metal-insulator-metal (MIM) capacitors, and should be much more stable. If the silicided layer, due to its metal-like properties, is assumed to be perfectly stable, then the dominant source of drift will be due to the underlying poly layer (Fig. 2.3), which is about 30× more resistive. As the drift of non-silicided poly resistors is between 0.2% and 0.8% (Chap. 2), the drift of silicided resistors should be 30× smaller, or no more than 300 ppm. This corresponds to an estimated WB temperature sensor drift of ~0.1 °C over its lifetime and thus a freuquency error of below 3 ppm for temperature-compensated MEMS oscillators. This hypothesis still needs to be verified by measurement results.

6.4.4 Energy-Efficient Wheatstone Bridge Temperature Sensors with Scaled Energy/Conversion

Although the WhB sensors presented in Chap. 4 can achieve state-of-the-art energy efficiency, their energy/conversion (~400 nJ/conv) is higher than most BJT-based sensors, while their sub-mK resolution is an overkill in most applications. For example, a battery-powered RFID sensor only requires a resolution of some tens of milli-Kelvin [31]. To extend battery life it would be desirable to trade resolution for energy/conversion along the 10 fJ·K^2 constant-FoM line (Fig. 6.1), for instance by achieving a thermal-noise-limited resolution of 10 mK while consuming 0.1 nJ/conversion (~4000× smaller). Lower energy/conversion can be achieved by reducing power consumption or shortening conversion time. Tuning the design parameters (e.g., resistor value, supply voltage, sampling frequency, ΔΣ-ADC order) could possibly achieve a 10× reduction. However, new architectures are required to obtain another 400× reduction.

To reduce the average power of the WhB front-end, heavy duty-cycling can be applied, and the WhB output voltage can be stored on a hold capacitor [32]. As for the readout circuit, it should provide sufficient resolution at a low oversampling ratio. This can be achieved by using an oversampled SAR ADC [33]. However, combining the two techniques could be challenging, as the hold capacitor in [32] cannot provide the low-impedance voltage input required by [33].

Another drawback of resistor-based sensors is the need for costly temperature calibrations. One way to simplify the calibration process is by integrating an inherently accurate temperature sensor, for example, a sensor based on an electro-thermal filter (ETF) on the same chip [6]. As a result, the sensor can be self-calibrated. Although the ETF will dissipate milli-watts of power, it can be disabled during normal operation, thus maintaining the sensor's energy efficiency. As mentioned in Sect. 6.4.2, integrating the two sensors with a (partially) shared readout circuit is necessary to reduce chip area and thus fabrication cost.

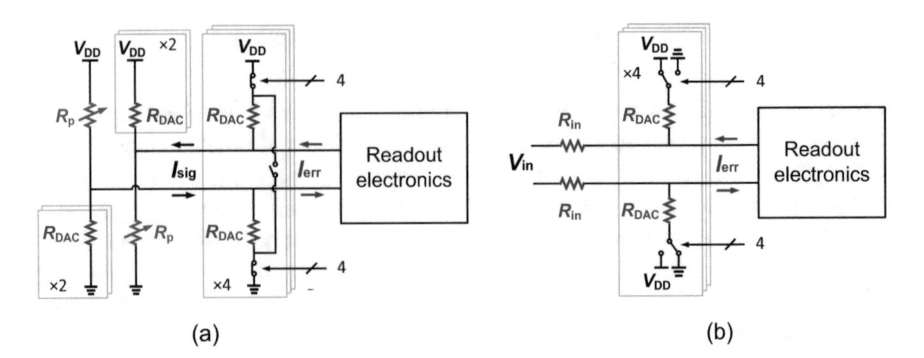

Fig. 6.6 (**a**) Wheatstone bridge front end shown in Sect. 4.6. (**b**) Input and DAC resistors of a CTΔΣ-ADC with equivalent input impedance and noise

6.4.5 Applications of the Tail-Resistor Linearized OTA

As shown in Chap. 4, Sect. 4.6, the nonlinearity of OTAs can be greatly improved by using the proposed tail-resistor linearization technique. As a result, the readout circuit of the Wheatstone bridge sensor can be scaled for thermal noise, which optimizes its efficiency.

The same principle can be applied to general-purpose CTΔΣ-ADCs, and Fig. 6.6 shows a simple translation while keeping the same readout electronics. By choosing the input resistor $R_{in} = R_p//(R_{DAC}/2) = 67\ k\Omega$ and switching the 4 DAC resistors (370 kΩ each) between supply rails, both the equivalent impedance and thus noise of the two topologies are the same. However, the DAC resistors in Fig. 6.6b consume less power than the Wheatstone bridge in Fig. 6.6a, and the estimated ADC power will be ~42 μW instead of 55 μW. Besides, the ADC will have a small stable input range, which is ~910 mV under a 1.8 V supply. Considering only its thermal noise, the ADC's expected Schreier FoM, which is DR+10·log(BW/power), should be ~183 dB. This is similar to the state-of-the-art CTΔΣ-ADC [34], and could be further improved by optimizing its input range, that is, the ratio between its input and DAC resistors.

Another attractive application of the linearized OTA is processing small biomedical signals of ~10 mV amplitude, for example, electromyography (EMG) signals. Because of its high linearity, the signal can be amplified with almost no distortion. Moreover, the OTA is capacitively coupled, so that it has a large input impedance and a large common-mode rejection ratio at DC, which are both advantageous in biomedical electronics.

6.5 Concluding Remarks

This book has discussed the development of energy-efficient resistor-based temperature sensors as well as their readout circuits. With this research, the energy efficiency of CMOS temperature sensors is improved by 65×, and is approaching the theoretical FoM limit set by the thermal noise of the sensor front-end (6×). Also, accuracy- and application-driven sensors have been developed.

However, there are still many directions left unexplored about resistor-based temperature sensors, and further improvements are required before commercializing the prototype sensors. Moreover, this book has also presented some general analog design techniques, for example, tail-resistor-based OTA linearization, which can potentially improve the performance of other circuits.

References

1. K.A.A. Makinwa, Smart temperature sensor survey, [Online]. Available: http://ei.ewi.tudelft.nl/docs/TSensor_survey.xls
2. C.H. Weng, C.K. Wu, T.H. Lin, A CMOS thermistor-embedded continuous-time delta-sigma temperature sensor with a resolution FoM of 0.65 pJ °C^2. IEEE J. Solid State Circuits **50**(11), 2491–2500 (2015)
3. M.H. Roshan et al., A MEMS-assisted temperature sensor with 20-μK resolution, conversion rate of 200 S/s, and FOM of 0.04 pJK^2. IEEE J. Solid State Circuits **52**(1), 185–197 (2017)
4. M. Cochet et al., A 225μm^2 probe single-point calibration digital temperature sensor using body-bias adjustment in 28 nm FD-SOI CMOS. IEEE Solid-State Circuits L. **1**(1), 14–17 (2018)
5. C. van Vroonhoven, D. D'Aquino, K. Makinwa, A ±0.4°C (3σ) −70 to 200°C time-domain temperature sensor based on heat diffusion in Si and SiO2, in *IEEE ISSCC Dig. Tech. Papers*, (2012), pp. 204–206
6. C.P.L. van Vroonhoven, D. d'Aquino, K.A.A. Makinwa, A thermal-diffusivity-based temperature sensor with an untrimmed inaccuracy of ±0.2°C (3σ) from −55°C to 125°C, in *IEEE ISSCC Dig. Tech. Papers*, (2010, Feb), pp. 314–315
7. B. Yousefzadeh, S.H. Shalmany, K.A.A. Makinwa, A BJT-based temperature-to-digital converter with ±60mK inaccuracy from −70°C to +125°C in 160nm CMOS. IEEE J. Solid State Circuits **52**(4), 1044–1052 (2017)
8. W. Zhao, R. Pan, Y. Ha, Z. Yang, A 0.4V 280-nW frequency reference-less nearly all-digital hybrid domain temperature sensor, in *IEEE Proc. ASSCC*, (2014, Nov), pp. 301–304
9. K. Pelzers, H. Xin, E. Cantatore, P. Harpe, A 2.18-pJ/conversion, 1656-μm^2 temperature sensor with a 0.61-pJ·K^2 FoM and 52-pW stand-by power. IEEE Solid-State Circuits L **3**, 82–85 (2020)
10. P. Chen, Y. Hu, J. Liou, B. Ren, A 486k S/s CMOS time-domain smart temperature sensor with −0.85°C/0.78°C voltage-calibrated error, in *Proc. ISCAS*, (2015, May), pp. 2109–2112
11. M.H. Perrott et al., A temperature-to-digital converter for a MEMS-based programmable oscillator with <±0.5-ppm frequency stability and <1-ps integrated jitter. IEEE J. Solid-State Circuits **48**(1), 276–291 (2013)
12. M.A. Pertijs, J.H. Huijsing, Charateristics of bipolar transistors, in *Precision Temperature Sensors in CMOS Technology*, (Springer, 2006)
13. M.S. Raman, T. Kifle, E. Bhattacharya, K.N. Bhat, Physical model for the resistivity and temperature coefficient of resistivity in heavily doped polysilicon. IEEE Trans. Electron Dev **53**(8), 1885–1892 (2006)
14. Y. Lee et al., A 5800-μm^2 resistor-based temperature sensor with a one-point trimmed inaccuracy of ±1.2°C (3σ) from −50°C to 105°C in 65-nm CMOS. Solid-State Circuits L. **2**(9), 67–70 (2019)
15. A. Khashaba et al., A 0.0088mm^2 resistor-based temperature sensor achieving 92fJ·K^2 FoM in 65nm CMOS, in *IEEE Dig. Tech. Papers*, (2020, Feb), pp. 60–61

16. W. Choi et al., A compact resistor-based CMOS temperature sensor with an inaccuracy of 0.12 °C (3σ) and a resolution FoM of 0.43 pJ·K^2 in 65-nm CMOS. IEEE J. Solid State Circuits **53**(12), 3356–3367 (2018)
17. S. Jeong, Z. Foo, Y. Lee, J. Sim, D. Blaauw, D. Sylvester, A fully-integrated 71 nW CMOS temperature sensor for low power wireless sensor nodes. IEEE J. Solid State Circuits **49**(8), 1682–1693 (2014)
18. X. Tang, K. Pun, W. Ng, A 0.9V 5kS/s resistor-based time-domain temperature sensor in 90nm CMOS with calibrated inaccuracy of −0.6°C/0.8°C from −40°C to 125°C, in *Proc. ASSCC*, (2013, Nov), pp. 169–172
19. Horng et al., A 0.7V resistive sensor with temperature/voltage detection function in 16nm FinFET technologies, in *IEEE Symp. VLSI Circ*, (2014, June), pp. 1–2
20. A. Mordakhay, J. Shor, Miniaturized, 0.01 mm^2, resistor-based thermal sensor with an energy consumption of 0.9 nJ and a conversion time of 80 μs for processor applications. IEEE J. Solid State Circuits **53**(10), 2958–2969 (2018)
21. H. Jiang, C.-C. Huang, M.R. Chan, D.A. Hall, A 2-in-1 temperature and humidity sensor with a single FLL Wheatstone-bridge front-end. IEEE J. Solid State Circuits **55**(8), 2174–2185 (2020)
22. J.A. Angevare, K.A.A. Makinwa, A 6800-μm^2 resistor-based temperature sensor with ±0.35 °C (3σ) inaccuracy in 180-nm CMOS. IEEE J. Solid State Circuits **54**(10), 2649–2657 (2019)
23. A. Wang, C. Chen, C. Liu, C.R. Shi, A 9-Bit resistor-based highly digital temperature sensor with a SAR-quantization embedded differential low-pass filter in 65-nm CMOS with a 2.5-μs conversion time. IEEE Sensors J. **19**(17), 7215–7225 (2019)
24. H. Xin, M. Andraud, P. Baltus, E. Cantatore, P. Harpe, A 0.34-571nW all-dynamic versatile sensor interface for temperature, capacitance, and resistance sensing, in *Proc. ESSCIRC*, (2019, Sept), pp. 161–164
25. H. Xin, M. Andraud, P. Baltus, E. Cantatore, P. Harpe, A 174 pW–488.3 nW 1 S/s–100 kS/s all-dynamic resistive temperature sensor with speed/resolution/resistance adaptability. IEEE Solid-State Circuits L. **1**(3), 70–73 (2018)
26. C. Wu, W. Chan, T. Lin, A 80kS/s 36μW resistor-based temperature sensor using BGR-free SAR ADC with a unevenly-weighted resistor string in 0.18μm CMOS, in *IEEE Symp. VLSI Circ*, (2011), pp. 222–223
27. H. Park, J. Kim, A 0.8-V resistor-based temperature sensor in 65-nm CMOS with supply sensitivity of 0.28 °C/V. IEEE J. Solid State Circuits **53**(3), 906–912 (2018)
28. Z. Tang, Y. Fang, X.-P. Yu, Z. Shi, L. Lin, N.N. Tan, A dynamic-biased resistor-based CMOS temperature sensor with a duty-cycle-modulated output. IEEE Trans. Circuits Syst. II **67**(9), 1504–1508 (2020)
29. J.A. Angevare, Y. Chae, K.A.A. Makinwa, A highly digital 2210μm2 resistor-based temperature sensor with a 1-point trimmed inaccuracy of ± 1.3 ° C (3 σ) from −55 ° C to 125 ° C in 65nm CMOS, in *IEEE ISSCC Dig. Tech. Papers*, (2021), pp. 76–78
30. G. Wang, A. Heidari, K.A.A. Makinwa, G.C.M. Meijer, An accurate BJT-based CMOS temperature sensor with duty-cycle-modulated output. IEEE Trans. Indus. Electron **64**(2), 1572–1580 (2017)
31. K. Souri, Y. Chae, K.A.A. Makinwa, A CMOS temperature sensor with a voltage-calibrated inaccuracy of ±0.15°C (3σ) from −55°C to 125°C. IEEE J. Solid State Circuits **48**(1), 292–301 (2013)
32. S. Oh et al., A 2.5nJ duty-cycled bridge-to-digital converter integrated in a 13mm^3 pressure-sensing system, in *IEEE ISSCC Dig. Tech. Papers*, (2018, Feb), pp. 328–330
33. Y. Shu, L. Kuo, T. Lo, An oversampling SAR ADC with DAC mismatch error shaping achieving 105 dB SFDR and 101 dB SNDR over 1 kHz BW in 55 nm CMOS. IEEE J. Solid State Circuits **51**(12), 2928–2940 (2016)
34. B. Gönen et al., A continuous-time zoom ADC for low-power audio applications. IEEE J. Solid State Circuits **55**(4), 1023–1031 (2020)

Appendix A

A.1 Measurement Setup

In this book, all the fabricated temperature sensors are characterized by a Pt-100 reference. The key of accurate characterization is creating a thermal equilibrium condition, so that the temperature error between the Pt-100 reference and the sensors can be minimized.

To achieve this, the chips were kept in a large aluminum box, with a customized lid to prevent airflow and thus temperature fluctuations, as shown in Fig. A.1. It also acted as a low-pass thermal filter with a low cut-of-frequency, making the sensor's temperature stable during measurements. For better thermal conductivity, the sensors were mounted inside the cavity of the aluminum box using thermal paste (Fig. A.2). The Pt-100 temperature reference was inserted into a hole in the metal block near the chips.

As shown in Fig. A.2, the Pt-100 is readout by an 8.5-digit Keithley 2002 multimeter. During temperature characterization, the bitstream capture would start only when the difference of two successive readouts of the Pt-100 temperature reference (about 20 s per readout) was less than 3 mK, held for about 400 successive readouts without any exception. Then the Pt-100 and the sensors could be considered in thermal equilibrium. With this setup, the oven settling time is about 4 hours for a single temperature point.

Moisture also plays an important role. As shown in Chap. 3, Sect. 3.4, it affects the stress profile of the plastic package and thus the value of the sensing resistor via the piezoresistive effect. Although this effect is much weaker with ceramic packages, it should be still avoided in order to achieve the best sensor inaccuracy. This can be done with dehumidifier, or simply by selecting the correct temperature ramping direction. As shown in Fig. A.3, when the test setup temperature is ramping down, the surface of the box is always cooler compared its cavity. Consequently,

S. Pan, K. A. A. Makinwa, *Resistor-based Temperature Sensors in CMOS Technology*, ACSP · Analog Circuits and Signal Processing,
https://doi.org/10.1007/978-3-030-95284-6

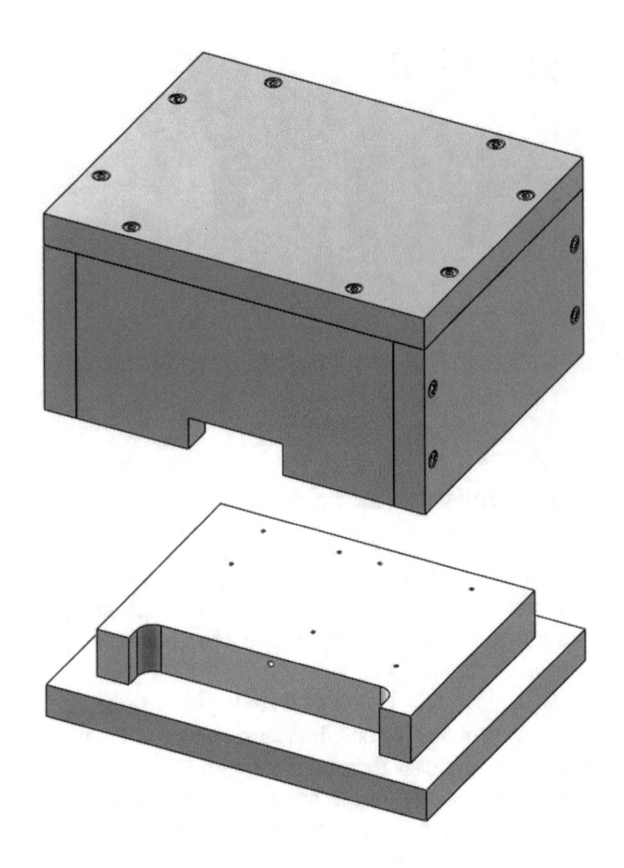

Fig. A.1 Customized metal box: base and lid

water droplets are generated on the surface of the metal box, but not the surface of packaged chips.

Last but not least, the noise of the clock reference should be minimized, especially for WB sensors. For the 1st WB sensor prototype (Chap. 3, Sect. 3.3), the reference clock was generated by a low-jitter (<1 ps) MEMS oscillator. For the rest of the experiments, it was generated by a function generator (Keysight 33600 A) with a similar jitter performance.

A.2 OTA with Tail-Resistor Linearization: Condition of the 3rd-Order Nonlinearity Cancellation

In this appendix, the optimum tail resistor value for the 3rd-order OTA nonlinearity suppression is derived, assuming that the transistors are operating in deep weak inversion.

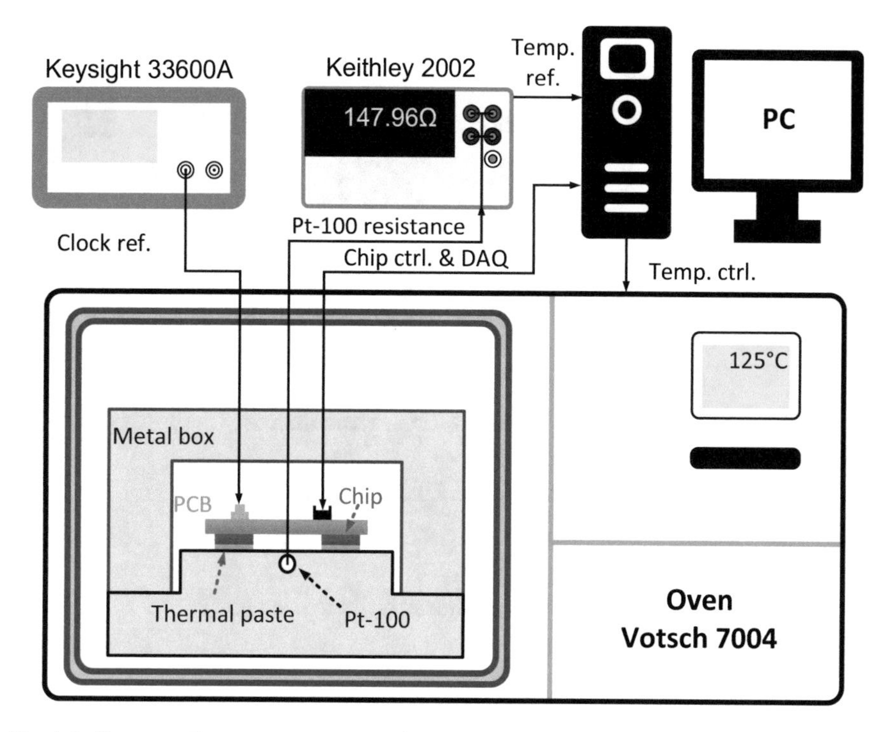

Fig. A.2 Test setup for temperature sensor characterization

For a single transistor in weak inversion region, its *V-I* characteristic is roughly exponential. After ignoring the effect of a finite output impedance and the back-gate effect, the output current I_d can be expressed as:

$$I_d = I_s \cdot e^{\frac{V_{gs} - V_T}{nU_T}}, \tag{A.1}$$

where V_T is the threshold voltage, I_s is the current at $V_{gs} = V_T$, n is a process-dependent slope factor, and $U_T = kT/q$ is the thermal voltage.

After applying a differential voltage $\pm\Delta V_g$ on the OTA with a tail resistor, the voltage drop on the tail resistor R_{tail} will increase due to the enlarged total current. Denoting the difference as ΔV_{tail}, the OTA's differential current output ΔI_o can be expressed as:

$$\Delta I_o = I_s \cdot e^{\frac{V_{gso} - V_T + \Delta V_g - \Delta V_{tail}}{nU_T}} - I_s \cdot e^{\frac{V_{gso} - V_T - \Delta V_g - \Delta V_{tail}}{nU_T}}. \tag{A.2}$$

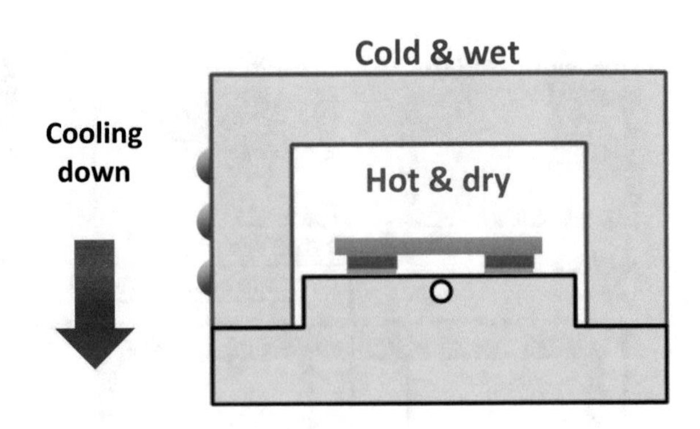

Fig. A.3 Moisture effect inside the metal box avoided by ramping down the oven temperature

Writing its Taylor series up to the 3rd-order term, there is

$$\Delta I_o = I_s \cdot e^{\frac{V_{gso}-V_T}{nU_T}} \left(\frac{2\Delta V_g}{nU_T} - \frac{4\Delta V_g \Delta V_{tail}}{2(nU_T)^2} + \frac{2\Delta V_g^{\,3} + 6\Delta V_g \Delta V_{tail}^{\,2}}{6(nU_T)^3} \right) \tag{A.3}$$

In order to cancel the 3rd-order nonlinearity, there must be

$$-\frac{4\Delta V_g \Delta V_{tail}}{2(nU_T)^2} + \frac{2\Delta V_g^{\,3}}{6(nU_T)^3} = 0. \tag{A.4}$$

Thus, the relationship between ΔV_g and ΔV_{tail} becomes

$$\Delta V_{tail} = \frac{\Delta V_g^{\,2}}{6nU_T}. \tag{A.5}$$

Alternatively, ΔV_{tail} can be calculated as $\Delta I_{tail} \cdot R_{tail}$, that is,

$$\Delta V_{tail} = R_{tail} \cdot I_s e^{\frac{V_{gso}-V_T}{nU_T}} \left(e^{\frac{\Delta V_g - \Delta V_s}{nU_T}} + e^{\frac{-\Delta V_g - \Delta V_s}{nU_T}} - 2 \right). \tag{A.6}$$

Writing its Tylor series up to the 2nd-order term, there is

$$\Delta V_{tail} = R_{tail} \cdot I_s e^{\frac{V_{gso}-V_T}{nU_T}} \left(\frac{-2\Delta V_{tail}}{nU_T} + \frac{2\Delta V_g^{\,2} + 2\Delta V_{tail}^{\,2}}{2(nU_T)^2} \right). \tag{A.7}$$

With $\Delta V_{tail} = \frac{\Delta V_g^{\,2}}{6nU_T}$, and neglecting the term with ΔV_{tail}^2, Eq. (A.7) can be rewritten as:

$$\frac{\Delta V_g^{\,2}}{6nU_T} = R_{tail} \cdot I_s \cdot e^{\frac{V_{gs0}-V_T}{nU_T}} \cdot \frac{2\Delta V_g^{\,2}}{3(nU_T)^2}. \tag{A.8}$$

Thus, the term ΔV_g^2 can be cancelled on both sides, leaving

$$R_{tail} = \frac{nU_T}{4I_s \cdot e^{\frac{V_{gso}-V_T}{nU_T}}} = \frac{nU_T}{2I_{tail}}, \tag{A.9}$$

where I_{tail} is the tail current of the differential pair given a zero differential input.

Index

S. Pan, K. A. A. Makinwa, *Resistor-based Temperature Sensors in CMOS Technology*, ACSP · Analog Circuits and Signal Processing,
https://doi.org/10.1007/978-3-030-95284-6

Printed in the United States
by Baker & Taylor Publisher Services